OPTICAL MEDIA

OPTICAL MEDIA:

Berlin Lectures 1999

FRIEDRICH KITTLER

Translated by Anthony Enns

polity

First published in German as *Optische Medien / Berliner Vorlesung 1999* © Merve Verlag Berlin, 2002

This English edition © Polity Press, 2010

Polity Press
65 Bridge Street
Cambridge CB2 1UR, UK

Polity Press
350 Main Street
Malden, MA 02148, USA

ISBN-13: 978-0-7456-4090-7 (hardback)
ISBN-13: 978-0-7456-4091-4 (paperback)

A catalogue record for this book is available from the British Library.

Typeset in 10.5 on 12 pt Sabon
by Toppan Best-set Premedia Limited

The publisher has used its best endeavors to ensure that the URLs for external websites referred to in this book are correct and active at the time of going to press. However, the publisher has no responsibility for the websites and can make no guarantee that a site will remain live or that the content is or will remain appropriate.

Every effort has been made to trace all copyright holders, but if any have been inadvertently overlooked the publisher will be pleased to include any necessary credits in any subsequent reprint or edition.

For further information on Polity, visit our website: www.politybooks.com

The translation of this work was supported by a grant from the Goethe-Institut that is funded by the Ministry of Foreign Affairs.

CONTENTS

INTRODUCTION: FRIEDRICH KITTLER'S LIGHT SHOWS

John Durham Peters

Optical Media may be Friedrich Kittler's best book for the uninitiated. It is breezy, has an off-the-cuff tone, and is generally free of the distinctive, pithy, and forbidding prose style that is sometimes called "Kittlerdeutsch." *Optical Media* provides an accessible introduction to the media theory of Kittler's middle period as it applies primarily to the optical realm. (Kittler's chief interest in sound, music, and above all time takes a back seat here.) It is definitely not "Kittler for Dummies," however. It assumes a minimal general knowledge of German literature and culture, the field in which Kittler began as a young renegade scholar in the 1970s, and it asks the reader to follow some ambitiously bold or bald claims. It presents the leading themes of Kittler's media theory and a few of its idiosyncrasies as well. The English-language reader can face a number of obstacles in a first meeting with Kittler, and this introduction seeks to ease the encounter.[1]

A Different Kind of German Import

Friedrich A. Kittler was born in 1943 in what was soon to become East Germany and lived there until 1958, when his family moved to West Germany. He studied German language and literature, Romance language and literature, and philosophy at Freiburg, where he became intimately familiar with recent French thought (he still claims to speak French more comfortably than English) and the work of Heidegger. In the 1980s and 1990s he was a visiting professor at several American universities such as Stanford, Berkeley, Santa Barbara, and Columbia. He has held the chair for "aesthetics and

1

media history" at Humboldt University in Berlin since 1993 and his fame and notoriety continue to grow. He is proud of the label given him by a supporter, "Europe's greatest media philosopher," and is sometimes called "the Derrida of the digital age" (although he would prefer to be the Foucault). Kittler is the pre-eminent theorist of elements – sound, image, letter, number, and organ – and how they are configured into media systems. He has produced a stunningly original and often controversial body of work that anyone interested in our fate as media-saturated animals today has to wrestle with.

Kittler is not in the same line as the leading German post-1968 imports into the Anglo-American human and social sciences. Specifically, his work should not be confused with hermeneutics or the Frankfurt School, which have been widely read in English, and it would be a bad mistake to see him as sharing their opinion of the degrading effects of technical rationality. Machines are our fate, according to Kittler, and to say so is not to witness to an awful downfall of the human condition; it is to properly grasp our situation.

Kittler takes great care in fact to differentiate himself from these two traditions. He is generally friendly to hermeneutics as a practice of rigorous textual analysis since it is close to his own method, but he mocks the spectral hope of communion with dead minds as a failure to recognize the material, i.e. medial, conditions of the practice of reading – conditions he analyzed at great length in his first blockbuster book, *Aufschreibesysteme*, translated as "Discourse Networks," and in a shorter collection called *Dichter, Mutter, Kind* (Poet, Mother, Child). This theme continues in these lectures, especially in his analysis of German romanticism.

Against the Frankfurt School, from Horkheimer and Adorno up through Habermas, however, Kittler is relentless in his scorn. Chief among many irritations is the notion of quantification as the disenchantment of the world. Kittler is a passionate friend of mathematics and an enemy of the notion that there is such a thing as "the two cultures" of science and humanities, as we say in English, or *Geist und Natur*, as they say in German. The idea that numbers rob our soul or that humanities have nothing to do with counting or machines he finds outrageous. (More fundamentally, he would find the claim outrageous that we have a soul to rob.) Music, dance, poetry – indeed all the human arts that involve time – would be nothing without counting, measure, and proportion. This is his current theme, pursued in his projected four-volume *magnum opus* on "Music and Mathematics." (The first volume, on "Ancient Greece," was published in 2006, and this turn to antiquity is one part of his work scarcely represented in *Optical Media*.[2]) Kittler blames the Frankfurt School for a willed blindness to our technical condition and a sentimentality about the humanities – and humans. To an outsider, the vehemence of his attacks seems out of proportion, especially since he shares with Adorno interests in inscription systems and technologies of sound, and an understanding of music as a field of play between excess and obedience. Some of the venom stems from turf wars in the German academy; some of it stems simply from his peevishness (something that the astute reader will note is not always absent from *Optical Media* either).

In any case, one will look in vain in Kittler's work for that reassuring dialectical defense of the human estate that many of us have come to expect in postwar German thought. Kittler is an altogether cooler and more ironic sort. Aesthetics he always understood in its original sense as sensation. To study aesthetics is not to study beauty *per se*, it is to study the materialities of our organs of perception. In this view he proudly belongs to the long Berlin tradition of psychophysical research into the human sense organs pioneered by figures such as Hermann Helmholtz. As *Optical Media* puts it in one of those reductive dicta for which Kittler is famous: "Aesthetic properties are always only dependent variables of technological feasibility." This is one reason the lectures are on optical and not visual media. Optics is a subfield of physics; vision is a subfield of physiology, psychology, and culture. The visible spectrum is a narrow band of a huge optical spectrum. For Kittler the subject is always subordinate to the object: human perception is an interface with physical realities. "Visual culture" is all the rage lately, but his focus on optics is a characteristic foregrounding of physical-technical conditions over perceived ones. The sense organs are signal processors, relatively weak ones at that, and Kittler rigorously refuses to take human quantities as the measure of all things; for him we cannot know our bodies and senses until they have been externalized in media. Although Kittler might be an outlier in terms of German imports, he is part of a tough-minded lineage within Germany that has not crossed the Channel or Atlantic as readily; the chief post-1968 exhibit would be the systems theory of Niklas Luhmann, which also takes vigorous distance from the Frankfurt School and the humanism of philosophical hermeneutics, and is an important source for Kittler and for all recent German-language media theory, which has turned out to be a wonderfully vital field of research and debate. Kittler is certainly not the only voice in German media theory. The publication in 1985 of *Aufschreibesysteme* was a watershed in the German humanities, and much in the subsequent flowering of media theory in the German-speaking world would probably be impossible without Kittler's innovations. His developing theories of media hardware had a huge influence in the 1990s, but he has been subject to sustained and incisive criticism and oedipal rebellion since in Germany. He is a controversialist who enjoys being outrageous and he has managed to offend or annoy a wide range of spirits. He is also brilliant and remarkably original. The task, as always, is to sort out the intellectual value from the rest. A prophet, as they say, has no honor in his own country, in part due to the sound ecological principle that any local species has co-evolved with its own natural enemies. In transplantation to an English-speaking universe, Kittler's work loses some of its outrageousness, in part because we are not invested in the local skirmishes. Derrida and Foucault were always bigger in the United States than in France; the lack of co-evolved enemies usually helps a crop proliferate.[3] But the enemies also know a thing or two and can save us the trouble of finding out for ourselves. Some classic complaints about Kittler in Germany – his politics, views on people in general and women in particular, and his treatment of historical evidence – will be noted here. Martin Heidegger certainly has crossed over into English, but here again it is important to get things right. Kittler takes Heidegger as the godfather of technical reflection and as his philosophical master. Heidegger has been read as a technophobe, a cultural pessimist lamenting technology's spell in the age of the world picture. Kittler, like many others, vigorously and persuasively contests this reading, taking Heidegger as the great thinker of technology. For Kittler's Heidegger, *technē* is our lot; he simply dared to think it most fully in all its inhuman implications. Heidegger's analyses are not a sob story but a sober disclosure of our condition. Once known as a young Turk who helped bring French poststructuralism to Germany, Kittler has more recently redescribed his career as a steady defense of Heidegger, simply reimporting his thought once it had been laundered offshore by thinkers such as Derrida and Foucault, who took on the task of surgically removing it from its historical embedment in the mess of National Socialism. Kittler's recent work embraces the Hellenophilic poetics and right-wing politics of the late Heidegger with a rather stunning enthusiasm. This means that Kittler is fully "out" as a German conservative, a creature that has no exact equivalent in British or American politics and seems to hold an enormous intellectual fascination for us, if the recent vogue in social and cultural theory of Carl Schmitt (and of course Heidegger) is any indication. Along with figures such as Derrida and Foucault, Herbert Marcuse and Jean-Paul Sartre,

4

Hannah Arendt, Emmanuel Levinas, and Leo Strauss, Jan Patočka and even Milan Kundera, Kittler is one key interpreter of the wildly diverse Heideggerian legacy. Compared to the hegemonic left-wing populism of Commonwealth media studies over the past decades, Kittler's vision is certainly a stark contrast.

This is not Cultural Studies

This is the second key obstacle: Kittler's disdain for people, or more specifically, for the category of experience. This is not to say that Kittler is a misanthrope – personally he can be quite cheerful and charming. But he has no use for the category of "the human" or "experience." He gives us a media studies without people. In a sense, Kittler is Mr. Anti-Cultural Studies. He rejects the entire empiricist tradition from Hobbes to Mill and beyond that is the background philosophy and intellectual orientation in the English-speaking world, with its liberal politics, its love of experience, its affirmation of agency, and its inductive method. (Although "empiricist" in media studies is sometimes used narrowly to mean quantitative research, the British empiricist tradition is the ultimate background philosophy for both cultural studies and social scientific research.) Kittler has no room for "the people" in either the Marxist or populist sense. In *Optical Media*, the only figures are inventors or artists, and he even makes fun of his own tendency to worship "bourgeois geniuses." Like the early Foucault, he is interested in historical ruptures and not the slow sedimentation of social change through everyday practices: he gives us evolution by jerks, not by creeps. He prefers to focus on turning points rather than the long state of play in between the drama. Agency Kittler tends to attribute to abstractions such as world war and not to living, breathing actors. He is not interested in audiences or effects, resistance or hegemony, stars or genres; he spends no time on subcultures, postcoloniality, gender, race, sexuality, ethnicity, or class. His account of television history is quite distinct from the key narratives in Anglo-American scholarship that explain the historical conditions and political settlements that made television full of miscellaneous "flow" (Raymond Williams) or tell how the box became integrated into a gendered domestic economy (David Morley, Lynn Spigel, and many others). Kittler is interested in the engineering. Just what kind of media studies is this?

At first glance, his vision of media studies hardly fits any of the dominant modes – and this is one of the things that makes it so bracing. Even so, there are moments in *Optical Media* where one can see a certain kind of political-economic analysis peeking through his account of the rise of the German film industry, and his use of literary texts might seem to point to an affinity with semiology, but even here, he never revels in the pleasure of the text as such, but always tries to boil down the text to an algorithm: he always has a larger explanatory (or debunking) agenda.

The one natural link of Kittler's work in extant media studies is to the Canadian tradition of Innis and McLuhan. Innis's great contribution was the notion of "bias" in time and space, and McLuhan's was the notion of media as human extensions (or amputations). Kittler's key contribution is the notion of "time-axis manipulation." He is the pre-eminent thinker of time-based media and what it means to edit the flow of time with technical means.[4] Like most scholars, Kittler would rather be compared to Innis than McLuhan and this is quite fair intellectually. Kittler's materialist account of history, love of ancient Greece, and disdain for the kind of body-humanism at the heart of McLuhan's thought puts him closer to Innis. Geoffrey Winthrop-Young, the most astute commentator on Kittler in English, calls Kittler "Innis in battle fatigues." But his persona – witty, arch, devil-may-care, politically incorrect – is closer to McLuhan. In tone, Kittler also is closer to McLuhan's flamboyant vaticism than to Innis's cranky accumulation of detail. Like both Canadians, his subject is the play of media in history, but he has taken a step forward over either. Over Innis's staccato inventories of historical occurrences and McLuhan's defiance of rigor, we have in Kittler a kind of media analysis whose method is dialectically acute and philosophically deep. Kittler brings a Hegelian ambition and lucidity to media studies. He begins *Optical Media* by frankly announcing his "insane and probably impossible" mission of doing for media what Hegel did for aesthetics. If Kittler does not fully succeed, he certainly goes further than anyone else in producing a philosophy of media and generating a rich array of productive ideas. (Hegel is, by the way, the first in a long line of German thinkers poorly understood in English; if Kittler is baffling to anglophone readers, he has distinguished company among his countrymen.) Is Kittler, then, a "technological determinist"? This slur is used freely and often without much nuance. If the question is, do humans have agency in the face of technology, then Kittler would be one, but not because he thinks technology is so determinant but, more radically, since he is not interested in what he persists in calling "so-called humans" (the joke is getting a bit stale). But if the question is, does he recognize the role of historical, political, and economic

6

contingency in the development of media, then he is not. Although Kittler is clearly a better media philosopher than media historian, and there are plenty of bones to pick with his historical interpretations, *Optical Media* is clearly interested in the development of institutions such as Edison's Lab, the ways that marketing imposes compromises between consumers and engineers, the unique historical conditions that enabled the emergence of photography in the 1830s, and the ways that war financing affected the development of television. His celebration elsewhere of the technological *bricoleur* who, with soldering gun in hand and DOS screen in view, reconfigures the user-friendly interfaces of dumbed down technologies, certainly suggests an activist role for at least those with an engineering bent. Indeed, in his 1990s campaign against Microsoft Windows, Kittler could sound a bit like a cultural studies type praising the agency of ordinary folks who rewrite the dominant code for their own purposes. This convergence on a DIY ethic in both cultural studies and Kittler points to a shared debt to punk culture. (At an even deeper level, Kittler and British cultural studies share a culturally Protestant love of anarchic textual interpretation by readers freed from the stupefying pictures of priestly authorities; E. P. Thompson's radical Methodists and Kittler's punk programmers might have a lot to say to each other.[5]) In the end, however, people alarmed by whatever they mean by "technological determinism" are likely to be alarmed by Kittler, especially his teleological claim, repeated throughout his work, that all media history culminates in the digital computer.

Then there is the question of gender. To say that people are simulations or effects of media conditions is a fine provocation that invites reflection on the moorings of human existence in an electrical (or any) age. Yet denying agency or subjectivity to people does not affect the two genders in the same way. For men, this might be a philosophically novel experience, but for women this is probably more of the same-old. The treatment of women in Kittler has always been problematic. *Discourse Networks* has remarkable insights about the role of mothers in producing romantic reading subjects – the mother serving as a kind of media relay between text and child, ear and eye, alphabet and voice – and *Gramophone, Film, Typewriter* has similarly stunning observations about how the typewriter desexualized the traditional division of labor in the act of writing. It is not that Kittler is not interested in gender; rather, it is that he is quite interested in sex. All of his media histories are also histories of eros, and this coupling of love and knowledge, music and media, and aphrodisiacs and mathematics is central to his story about the

7

ancient Greeks; *Music and Mathematics* might be read as an alternate take on Foucault's own multi-volume history of sexuality (without the queer subtext). His books and lectures are marked by occasional bawdy comments (not to mention pictures), something that reveals an unreconstructed gender politics that would be hard to find surfacing openly in a university in the US or UK in the past two decades. (It is not hard to imagine it surfacing in politics, sports, or show business, however; on this score, Kittler is, as he once said, "stinkingly ordinary.") In Kittler's work, the default gender roles often have men as soldiers and women as pin-ups. *Caveat lector*. Kittler all but invites a gender-sensitive reading, and his work reminds us of the feminist adage that gender and technology are not separately constituted categories. The first volume of his *Music and Mathematics* is full of praise for women – Aphrodite, Circe, Sappho – but they are all goddesses or legends. Ordinary ones who don't partake of what Goethe famously called the "eternal feminine" don't fare so well in his work. Here again, ordinary people of any sort don't fare well, but the grievance is clearer on one side of the divide.

Scholarly Extremities

Robert Boyle, the great experimentalist of the seventeenth century, famously invented a new kind of writing, the scientific report. In this genre, cognitive virtue was tied to a public demonstration of methods that anyone could repeat and an expository style that anyone could understand. Readers of English scholarly prose still tend to expect a basic level of reasonableness and clarity – expectations that Kittler, like most other high-flying German thinkers since Hegel – will regularly topple. His conclusions leap like lightning. He argues by anecdote rather than induction, and will always choose a dramatic narrative line over one that muddles through the complexities on the ground. He loves the trope of synecdoche, the part for the whole, and wants single instances to resonate with unspoken richness. We are meant to read the title of his book, *Gramophone, Film, Typewriter*, for instance, as standing in for modern acoustic, optical, and word-processing techniques in general. Like anyone schooled in modern literature, he is fond of allusion. At key points in *Optical Media*, he does not name things that he thinks should be obvious, managing to avoid mentioning "the uncanny" in his discussion of Freud's encounter with his mirror image on a train or "fractals" in his discussion of Mandelbrot's computer graphics. We are supposed to be in the

8

know. He also can't resist a lame joke, an irreverent comment, or a prodding exaggeration. Thus, Ronald Reagan is "emperor of California," America "the country of unlimited serialism," democracy the "age of prevailing illiteracy," and poetry has been tedious for 2,000 years. The most charitable way to read his style is that he treats his audience at a high (i.e. his) level, and follows the Heideggerian principle that names stop thinking. Less charitable readers can find the style arrogant or obscurantist. At least Kittler is never boring and always strives to see things afresh. Aristotle, Hegel, Mead, and Wittgenstein are among the many philosophers whose reputations profited from posthumously published lectures. Publishing lectures while one is alive changes the equation a bit. These lectures have a liberty of citation and factual reference that might be more forgivable as oral utterance or student transcriptions of a beloved teacher now deceased. Kittler's critics often have beefs with the details of his scholarship, and his tendency to argue in broad strokes is even more evident in these lectures than usual. Film scholars have found much to object to in his treatment of silent cinema, for instance – indeed, just as they did to his section on cinema in *Gramophone, Film, Typewriter*.[6] In fairness, he admits a lack of evidence for some speculations in *Optical Media* and says his treatment of film history is full of "rough cuts." But on the other hand, Kittler himself is not always fair: he is unjustly snippy about both Friedrich von Zglinicki, on whose work he heavily relies, and the academic field of film studies, which he mocks as trading in "cultural-history gossip." A committed intellectual poacher and interdisciplinarian whose interests range from the history of mathematics to German poetry, from computer programming to warfare, from classical Greek philology to psychoanalysis, Kittler can be impatient in conforming to the standards of academic specialism. He can share something of McLuhan's stance of writing off whole areas of study as missing the boat, as simply not "getting it." (The boat, of course, is one whose rudder both men steer, the S. S. Media.)

How damaging are Kittler's flippant tone and sometimes cavalier stance to scholarly norms? Scholarship is governed by a diversity of values such as accuracy, excitement, judgment, novelty, and fairness. Clearly, Kittler scores better on some of these than others. But there is plenty of room in the academic ecosystem for different sorts of contributions, and no one scores a perfect ten in everything. The German university has a long tradition of providing a platform for stylistically wild and scandalously ambitious intellectual claims; it also has a long tradition of tediously thorough toil over detail. It is no secret that Kittler belongs to the first tradition, which he supplements with pyrotechnic methods learned from the French. Of course, it is crucial to take Kittler, like everyone else, with a pinch of salt. But if all you see are the mistakes and exaggerations, you risk missing something big if you are interested in understanding media as architectonically important to our life and times. Professional historians find Foucault wrong on many things: but that does not make him any less interesting or important; it leaves more work to do. (Kittler probably covers his tracks less well than Foucault.) Many a great thinker has survived an occasional howler; indeed, producing howlers may be part of the job. It takes a first-rate thinker, as Hannah Arendt noted, to produce a first-rate contradiction. In pointing out Kittler's problems, I sometimes worry that we critics risk standing on his shoulders and punching him on the ears. Scholars, like all people, measure others by their own size. Prophets are a rarer species than scribes. It is a whole lot easier to show where another scholar went off the rails than to invent a new style of thought. There is a growing body of sound and interesting scholarship on media history in all its flavors; but there is very little thinking that is as breathtakingly imaginative as Kittler's. His mammoth work comes with some nasty little gnats flying around it, but we shouldn't let the gnats blind us to the magnificence of the shaggy beast he has brought on to the scene.

Context and Overview of *Optical Media*

Optical Media distills many of the theses of *Discourse Networks* and *Gramophone, Film, Typewriter*. It is not meant to be the presentation of cutting-edge research; it is a kind of popular exposition of a body of work. The lecture, as Erving Goffman points out, is a form of talk that is in cahoots with what he calls the "cognitive establishment."[7] Every lecture ultimately makes the promise that the world can be made sense of, and Kittler does not fail in his end of the bargain. Though much has rightly been made of his links to French poststructuralism (a term he dismisses), Kittler has none of the epistemological hypochondria that some Anglo-American followers – especially of Derrida – take away from the movement. Kittler has no interest in "interrogating" our "responsibility" for the "aporias" of "undecidability" (to "deploy" several deconstructionist clichés); he charges full steam ahead with the business of enlightenment. Wielding a scalpel of Hegelian sharpness Kittler confidently slices truth from nonsense. Unlike some practitioners of cultural studies, he seems unembarrassed to be a professor, although he always discusses that role with a sly irony. Indeed, one context for *Optical Media* is the establishment of media studies at Humboldt University, which, as the former University of Berlin, is the ground zero of the modern research university. Kittler's book from the same period, *Eine Kulturgeschichte der Kulturwissenschaft* (2001), is a lecture series on the intellectual history of university life; in *Optical Media*, the lecture as a medium is always a topic for self-reflection along with running in-jokes about professors and universities. *Optical Media* does not agonize about the inherent terror of reason; it gets on with the task of dispensing knowledge to students.

The lectures are relatively easy to follow and do not require any extensive summary; here I simply sketch the narrative arc. Kittler's exposition follows a more or less threefold narrative of artistic, analog, and digital media. The history is quite conventional in its basic periodization – Renaissance, Reformation, Counter-Reformation, Enlightenment, Romanticism, etc. The revisionism occurs, rather, inside the well-worn categories as Kittler insinuates media into them, showing how much media explain what we think we already know. The era of artistic media is governed by the hand and reaches its peak in the linear perspective of the Renaissance. This is not a hand innocent of technique, but rather one tutored by perspective's discipline of geometry or the inverted tracings of light itself in the *camera obscura*. As always, these artistic media are in cahoots with state, religious, and military power (especially the magic lantern and other techniques of projection). The era of analog media – that of optical media proper – frees the act of visual depiction from the human hand and the act of visual perception from the human eye. A series of photographic devices allows for a kind of direct transcription of the sunshine without intervention of the pencil or brush, and liberates the realm of the visible from the physiology of the eye. The human being is no longer the lord of the record or of the knowable universe. Machines take on tasks – drawing, writing, seeing, hearing, word-processing, memory, and even knowing – that once were thought unique to humans and often perform them better. (That the "uniquely human" recedes before the onslaught of new media is one of Kittler's characteristic claims.) The great breakthrough of analog media is the storage and manipulation of temporal process; the great problem is the lack of interoperability between systems. Kittler's account of cinema and television is a rather teleological story about how the optical and acoustical tracks got patched together in various configurations along the road to compatibility and convertibility. That great apotheosis had to await the digital moment. Now we await the utopian possibility, as he concludes, of working with light as light, not just light as effect or trace. There is something almost millenarian in this culmination of media history. The end of history returns us to the beginning, when the light was separated from the darkness.

The Love of Light

What does require comment are the ways these lectures act out a number of key themes and gestures of Kittler's larger media theory. Seven is a good number:

1. Abstraction

Kittler makes clear that his subject is optical media and not, say, film and television history. He does not want to identify media with any particular incarnation and places "general principles of image storage, transmission, and processing above their various realizations." More specifically, the computer provides Kittler with a handy device for rewriting the history of all hitherto existing media. Media are data-processors: this is his starting point. These lectures foreground the primacy of telecommunications – from the telegraph to the computer – as the foundation of modern optical media, along with the accompanying three functions of storage, transmission, and processing. Kittler's abstraction may shave off the edges of historical complexity, but he is always first and foremost a philosopher of media history rather than a media historian. His aim is to use history to inform philosophical reflection about techniques of sending, saving, and calculating. The *camera obscura* receives images; the magic lantern sends them; the camera stores them. Kittler loves rigorous schemas, especially if they have three parts.

2. Analogy

Media studies has a long history of showy displays of miscellaneous reference and magpie learning. One of McLuhan's favorite tricks was to create a surrealist juxtaposition of two distinct historical items, one usually highbrow and canonic, the other usually modern and popular. We see something similar in Kittler's claims that Baroque candles illuminate Sunday night talk shows, the positive-negative process in

12

photography is an early Boolean circuit or transistor, or poetry and dance, due to their basic periodicity, are susceptible to cylindrical storage devices like the phonograph. The first requirement for any good structuralist is a gift for discovering surprising categories and stunning analogies, and Kittler has this gift in spades. The greatest of all historical analogists, Hegel, frankly admitted that every analogy is wrong in some way, and Kittler shares with Hegel a delight in "the negative," precisely the part that does not fit and thus forces thought forward. This is part of his willingness to make insightful mistakes. Yet many of Kittler's analogies are clearly brilliant. His point that in 1840 the letter is to the telegraph what the painting is to photography clearly captures the epochal shift between manual and machine writing. Wagner as early cinema is abundantly suggestive. Taking Renaissance perspective as a variant on ballistics is smashingly good, even if it is slightly overcalculated to upset humanist pieties.

3. Writing

Literary study is the historical nexus of media studies in Britain, Canada, France, Germany, and Italy (but not the United States). The difference is that German philology for important historical reasons always took itself seriously as Wissenschaft, as a contributor to knowledge, whereas English literary criticism has often (not always) been happy to be edifying, an enhancement of sensibility rather than a kind of science. Kittler not only brings the full force of the German philosophical tradition to media studies; he also brings the full resources of the German philological tradition. Kittler was trained as a philologist, and a papyriological sleuthing through an archive of documents remains his central method. Media philology, as conducted by Kittler, starts from the fundamental fact of writing – which he takes to be the mother of all media. Writing should be understood in the expanded sense of inscription – a notion Kittler got equally from the French obsession with the practice of *écriture* as from computer lingo about writable and readable storage. Photograph, phonograph, telegraph – the threefold historical mutation of light, sound, and word-processing in the nineteenth century explored in *Gramophone, Film, Typewriter* – all pay implicit tribute to writing in their names as do numerous other electrical media practices. Kittler's skill at reading what was never written is at the core of his media analysis.

One of his key concepts is "the monopoly of writing" (note the Innisian accent). Before the nineteenth century, the only possible

13

form of cultural storage was the watery stuff of memory or writing. Writing had a monopoly on recording. The phonograph and film break this grip, since both allow for the writing of new kinds of material – sound and light. More specifically, they allow for the recording and thus manipulation, of time in its flow. Slow-motion, time-lapse, back-masking, jump-cuts, collage – modern acoustic and optical media – make temporal events available for editing. Kittler, again, takes the ability to play with the time axis as definitive of technical media. Even writing is a kind of incursion into time (storing words for later), though in a less radical way than film or phonograph.

4. Reading

Kittler is also a theorist of reading, especially of how what Habermas calls the "bourgeois public sphere" came into being, although Kittler's analysis is opposed in every way to Habermas's. Kittler's original analysis of Romantic literature is on full display in *Optical Media*. Romanticism for Kittler is not an attitude towards love, yearning, death, or distance; it is a particular use of book technology. Here, Kittler is in full arch debunker mode as he reprises some basic arguments of *Discourse Networks*. Reading is an "inner hallucination" by which the reader, who learned the technique on his or her mother's lap, decodes the text into a stream of sounds and images, even smells and tastes. Romantic poetry is the last gasp of the monopoly of writing before its nineteenth-century meltdown into audiovisual media. After that, writing could no longer bear the burden of holding sound and image, rival media having stripped it of those functions. Romantic reading is a kind of proto-film viewing, a lonely egotistical position within a simulating apparatus, not unlike the position of the fighter pilot or the video game player. Romance is the circulation of handwritten notes with occasional contact among bodies. In one of his dourly wonderful quips, Kittler notes that there can be media techniques without love, but no love without media techniques. Although he always disdains the humanist body that is so dutifully governed by its mind, Kittler always returns to the erotic body. The erotic, for Kittler, is always the first register of a media innovation. With his interest in excess and intoxication we find a window to a certain strange kind of agency. The most radical of all his claims is to reread all human organs as kinds of media apparatus; these organs are not only the eyes and ears. In *Music and Mathematics*, he sees the Greeks as the first explorers of the erotic effects of a new media regime (the alphabet in their case), which provides yet another way of making the ancient Greeks our contemporaries.

5. The Medium is the Message

Media for Kittler are always about themselves. In this, he subordinates content to form, quite in the spirit of McLuhan. Thus, the *Hörspiel* (or German radio drama) is about the conditions of radio listening. Cinema is about flying or seeing. He nicely begins *Optical Media* with a reflection on the medium of the lecture – a system made up of text and voice on the sending end, ears and half as many hands on the receiving end, and not (yet) interrupted by visual media. (His lectures were blessedly free of PowerPoint, a bacillus that has now all but entirely colonized the habitat.) Kittler is aggressively formalist in his analysis, and enjoys inducing the defamiliarizing shock of not caring about what a medium says. He is interested above all in a kind of engineering analysis of carrying capacities. Art has style, media have standards; artistic reproduction yields similar versions, while media copies are identical; art cannot inscribe evidentiary detail, while media are forensic and provide data for witnessing; and art is less easily automated than media (this is not necessarily praise for art, of course). Here lies a whole collection of architectonically crucial distinctions.

6. War

Kittler has recently said that after studying war for so long, he has decided to study love instead. This turn is clear in *Music and Mathematics*. But in *Optical Media*, war is still the mother of all things.[8] He gets remarkable mileage out of the military context of media in these lectures. In general, I find Kittler's emphasis on war a useful antidote to a myopically civilian approach that has marked most of our media histories. He produces lots of cool connections – celluloid as a kind of plastic explosive, the Counter-Reformation as a media battle, or the historic importance of blinding in both battlefield and ballroom, bedazzlement being both an aristocratic privilege and military tactic. The idea that the trenches of World War I were a kind of media laboratory for experimenting with the first mass audience of guinea-pigs/conscripts or that war itself has increasingly become a kind of media simulation are key Kittlerian points developed at greater length in *Gramophone, Film, Typewriter*. His notion of "the arming of the eye" owes something to Paul Virilio and is certainly relevant in an age of camera-guided missiles. Next time

15

you have your eyes checked, your ophthalmologist may ask you to "track" a "target." But all good analogies can get pushed too far. Of Kittler's link of the Colt revolver and film, Frank Kessler puckishly asks whether the sewing machine wouldn't serve Kittler as a better harbinger of serial processing, knowing that he would never warm to such a lowly, unwarlike domestic device. Kittler's fascination with war can sometimes seem slightly unhealthy, but there is no doubt that media history without the military-industrial complex is ultimately deeply misguided.

7. Light

Optical Media begins by praising the sun – a basic and brilliant fact before us that none of us directly see – courtesy of Dante and Leonardo. The sun is the condition of all seeing. It is a medium: we do not see it, but we see everything by way of it. (Media take the instrumental or ablative case: they are things by which something occurs.) The sun is both obvious and profound, and this beginning features Kittler at his most elemental, in his guise as a devotee of Mediterranean light, a celebrant of illumination and its intoxication. (It also reprises, in a curious way, McLuhan's claim about the electric light as an arch-medium.) If the eye is the light of the body, then the great star – the sun – as Dante says, is the light of the intelligence. In the end, what I like best about Kittler is his sheer love of intelligence and his commitment to delirious delight as a path to higher wisdom. Like all of us, Friedrich Kittler can be blind, but like very few of us, he can also be absolutely dazzling.[9]

Notes

1   The best introductions to Kittler's work in English are Geoffrey Winthrop-Young and Michael Wutz, "Translator's Introduction: Friedrich Kittler and Media Discourse Analysis," in *Gramophone, Film, Typewriter* (Stanford: Stanford University Press, 1999), pp. xi–xxxviii, and the special issue of *Theory, Culture & Society* 23 (2006), nos. 7–8. I have not attempted to provide a full listing of sources on Kittler in English in these notes.
2   For a more sustained discussion of this turn, see Claudia Breger, "Gods, German Scholars, and the Gift of Greece: Friedrich Kittler's Philhellenic Fantasies," *Theory, Culture & Society* 23 (2006): 111–34.

3   See my "Strange Sympathies: Horizons of Media Theory in America and Germany," *American Studies as Media Studies*, eds. Frank Kelleter and Daniel Stein (Heidelberg: Universitätsverlag Winter, 2008), pp. 3–23.

4   Sybille Krämer, "The Cultural Techniques of Time Axis Manipulation: On Friedrich Kittler's Conception of Media," *Theory, Culture & Society* 23 (2006): 93–109.

5   Geoffrey Winthrop-Young, "Silicon Sociology, or Two Kings on Hegel's Throne? Kittler, Luhmann, and the Posthuman Merger of German Media Theory," *Yale Journal of Criticism* 13 (2000): 391–420.

6   See Frank Kessler, "Bilder in Bewegung: Für eine nicht-teleologische Mediengeschichtsschreibung," *Apparaturen bewegter Bilder*, ed. Daniel Gethmann and Christoph B. Schulz (Munich: Lit Verlag, 2006), pp. 208–20, and "Medienhistorische Erleuchtungen," *KINtop: Jahrbuch zur Erforschung des frühen Films* 13 (2005): 177–9.

7   Erving Goffman, "The Lecture," *Forms of Talk* (Philadelphia: University of Pennsylvania Press, 1981), pp. 160–96; see especially p. 195.

8   See Geoffrey Winthrop-Young, "Drill and Distraction in the Yellow Submarine: On the Dominance of War in Friedrich Kittler's Media Theory," *Critical Inquiry* 28 (2002): 825–54.

9   For helpful commentary I would like to thank Gina Giotta, Klaus Bruhn Jensen, Frank Kessler, Benjamin Peters, John Thompson, and Geoffrey Winthrop-Young, without incriminating them in any of my opinions.

ACKNOWLEDGEMENTS

We are grateful to the following for permission to reproduce copyright material:

Professor Giuseppa Saccaro Del Buffa, Battisti for an extract from *Brunelleschi: The Complete Works* by Eugenio Battisti, 1981, originally published by Electa Editrice, reproduced with permission; Georg Olms Verlag AG for an extract from *Der Weg des Films* Friedrich von Zglinicki, 1979 copyright © Georg Olms Verlag AG; Piper Verlag GmbH for 14 words from the poem "To the Sun" ("An die Sonne") by Ingeborg Bachmann, first published in "Merkur. Deutsche Zeitschrift für europäisches Denken." Vol. 10, Nr. 6, June 1956, p. 534. Copyright Piper Verlag: included in a collection of poems called "Anrufung des Großen Bären"; Suhrkamp Insel 10 words from Hans Magnus Enzensberger's poem "J.G.G. (1395–1468)". First published in a collection of poems called "Mausoleum. Siebenunddreißig Balladen aus der Geschichte des Fortschritts" ("Mausoleum. Thirty-seven ballads from the history of progress") Frankfurt / Main, 1975, p. 9 copyright Suhrkamp Verlag.

In some instances we have been unable to trace the owners of copyright material and we would appreciate any information that would enable us to do so.

PREFACE

O gloriose stelle, o lume pregno
Di gran virtu, dal quale io riconosco
Tutto, che si sia, lo mio ingegno.
Dante, *Paradiso* XXII

To be purely truthful, every lecture concerning optical media should begin by praising the star that first made it possible to see earthly things at all. "There is nothing more beautiful under the sun than to be under the sun," wrote Ingeborg Bachmann from the humble viewpoint of this Earth (Bachmann, 1994, p. 219). Leonardo da Vinci, an older and more arrogant European, said the same thing from the viewpoint of the sun itself: "Il sole non vide mai nessuna ombra – The sun never sees a shadow." (Codex Atlanticus, 300 r.b.)

But in a world whose everyday life is determined by science and technology rather than the sun, lectures are always already on the other side of light. The optical media in my title all act and operate in that shadow, which the sun, according to Leonardo, does not see. In other words, art and technology represent two different ways of shifting the boundaries of visuality, so to speak, by either misusing or circumventing the sun. At first, these lectures will look at European painting since the Renaissance in a traditional or aesthetic way, in order to discern the principles according to which modern visual perception was organized. In this artisanal, hand-crafted phase of optical media, however, it should already become clear that they would not have been conceivable without calculations, and thus they also required a technical-scientific foundation. The technical apparatus could therefore detach itself from the eyes and hands of so-called artists and form those absolutely autonomous spheres – optical media

19

technologies – that surround us or even determine us today. The path of the lecture leads, to sum it up in one sentence, from Renaissance linear perspective, past the almost already old-fashioned technologies of photography, film, and television, to late twentieth-century computer graphics.

In doing so, however, I am taking on an apparently impossible task: I am using texts to speak about relatively recent lived realities, which are by definition neither language nor text, neither oral nor written. Photographs, films, and television screens normally have no place in the humanities. Indeed, they had no place in academic lectures at all, no matter what the discipline, so long as universities were universities, meaning that the German state was not yet committed to installing audiovisual technology at precisely the same time in courts, prisons, and traffic intersections for police surveillance, as well as in academic lecture halls. The state agreed to supply video cameras and monitors – I am citing from official documents – as "a necessary technical adaptation of public education institutions to the communication level of the times and its financial, organizational, and politico-educational effects." This is how it came about that the great art historian Heinrich Wölfflin – according to a comment by Horst Bredekamp – attributed his greatness above all to the fact that Wölfflin or one of his assistants invented the dual projection of all the images dealt with in his lectures.

Before this technical break, on the other hand, lectures were not visual at all. One hundred and twenty years ago, as a young philosophy professor in Basel, Friedrich Nietzsche described how classical German universities functioned. In the last of his five lectures entitled *On the Future of Our Educational Institutions* he writes:

> If a foreigner desires to know something of the methods of our universities, he asks first of all with emphasis: "How is the student connected with the university?" We answer: "By the ear, as a hearer." The foreigner is astonished. "Only by the ear?" he repeats. "Only by the ear," we again reply. The student hears. When he speaks, when he sees, when he is in the company of his companions, when he takes up some branch of art: in short, when he lives, he is independent, i.e. not dependent upon the educational institution. The student very often writes down something while he hears; and it is only at these rare moments that he hangs to the umbilical cord of his Alma Mater. He himself may choose what he is to listen to; he is not bound to believe what is said; he may close his ears if he does not care to hear [. . .] The teacher, however, speaks to these listening students. Whatever else he may think and do is cut off from the student's perception by

20

an immense gap. The professor often reads when he is speaking. As a rule he wishes to have as many hearers as possible; he is not content to have a few, and he is never satisfied with one only. One speaking mouth, with many ears, and half as many writing hands – there you have to all appearances, the external academical apparatus; the university engine of culture set in motion. (Nietzsche, 1964, pp. 125–6)

This is Nietzsche's incredibly precise, because it is ethnological, description of our workplace, whose audiovisual future he could not yet have foreseen. As you can see, or rather hear, the absence of visual pleasure in the old universities was no small source of pride. A lecture, therefore, via the ear, or eventually even on the radio, remained totally within the classical framework; the realm of the old European universities wasn't conclusively blown open until optical media, i.e. that which "the student" (and since the Prussian educational reforms of 1908 this also included the female student) sees, became a subject.

I confess, however, that all of Nietzsche's irony cannot bring me to elevate the theme of these lectures to its medium. In other words, I will make no use of the state's offer of video recorders, monitors, and projectors for pedagogical or other purposes. If it turns out to be possible – that is, above all, if someone manages to find an interface between computer and television monitors during the course of the semester – visual examples should be woven in, but I would prefer the experimental rather than the entertaining, silent film or computer graphics rather than blockbusters. Otherwise, the medium of the lecture will generally remain the same mixture of acoustics and textuality that Nietzsche so ironically and precisely described. This at least has the methodological advantage that it looks at contemporary optical media from the exact same outside and ethnological perspective that Nietzsche applied to the activity of lecturing in his own day.

I thus come to the question of the subject matter and methods that you should expect here. In order not to disappoint latecomers (like myself this semester), the remainder of today's general lecture will discuss first the subject matter, second its possible practical relevance, and third the methods.

Concerning the subject matter – as I said, the lecture will convey an ethnological look at the wealth of man-made images of the last hundred years, and it will therefore precisely fill in the time between Nietzsche's diagnosis and the present. To do that, though, we must first go further afield, and the first step of all is to tackle the long prehistory of contemporary optical media, in which images were

21

actually painted but could neither be stored nor transmitted, let alone move. In this prehistory, images appeared together with literary texts – as book illustrations or diagrams, as pictures of mythological models, or finally as imaginary images produced by literature in the so-called inner eye of the reader. In this respect, there were centuries at least that already belonged to the history of optical media, as they dreamed of modern technologies and developed mechanical devices, whose scientific realization was finally made possible in the nineteenth century through photography and film. The most significant among these apparatuses are above all the *camera obscura* as a device for recording images and the *lanterna magica* as a device for reproducing images. These devices were more than mere technical aids for perspectival painting since the time of the Renaissance, and their connections to the basic technology of the printing press, as well as the role they performed in the media war between the Reformation and the Counter-Reformation or between the printing press and church imagery, will be thoroughly dealt with. For film did not fall from heaven, but rather it can only be understood through the fantasies and the politics that its invention was responding to. The fact that television, as far as I can see, was not once seriously imagined until its factual development also calls for analysis.

In the second step, after a run-through of the prehistory, we will examine the history of how images first learned to be stored and then also to move. As we know, both happened in the nineteenth century, which began with the development of photography and ended with the development of film. To understand this long nineteenth century, as Martin Heidegger called it, media historically we must address the question of how the new image technologies especially affected the old arts, which had been handed down for centuries or millennia. The competitive relationship between photography and painting is well known, but less is known about the relationship between film and theater. With the exception of a single theater historian, little light has been shed on how ballet, opera, and theater – at least since the nineteenth century, but also in innovations like the baroque proscenium or "picture-frame" stage – evolved from elements that would later constitute cinema. This can be seen above all in Babbage and Faraday's lighting engineering, but it was epitomized in Wagner's *Gesamtkunstwerk*. Proof should therefore be offered that Wagnerian opera is really and truly cinematic, as Nietzsche's *The Birth of Tragedy* already predicted.

The local aim of such arguments is also to prove clearly that Humboldt University is fully justified in joining the disciplines of media history, art history, musicology, and theater studies into a single faculty. Central is the question of the effects that the development of film in particular had on the ancient monopoly of writing itself. Theater and opera are only examples of art forms that functioned in parasitical dependence on the monopoly of writing; think about the role of role in theater or score in opera. On the other hand, the introduction of technical media disrupts this monopoly as such, and it therefore works on a level that is more radical than competition. The issue thus becomes not only which forms of competition film provoked for novelists – since 1910 they are all somewhere between the extremes of having their work adapted or rejected by the film industry – but also the new status of books themselves under audiovisual conditions. The range of possibilities is quite diverse: there are books whose letters form themselves into images, other books that are written such that they can be hallucinated as films, and yet others (like Kafka's story *The Judgment*) that refuse every illustration. And when one considers that contemporary novelists like Thomas Pynchon (who will play the role of a principal witness for film and television in these lectures) have had all photographs of themselves destroyed, it is possible to measure the abyss between literature and optical media.

But to come back to the media themselves: in the third step, these lectures will attempt to derive the structures of film and cinema from the history of their development. In the sequence from silent film to sound and color film – three stages that oddly enough correlate fairly closely with the outbreak of two world wars – we see the emergence of different media-specific solutions (if not outright tricks). On the one hand, I will attempt to present these solutions in a technical way in order to incorporate what film analysis and film semiotics normally have to teach concerning montage, focus, lighting, directing, etc. Elementary facts concerning the film material, the film apparatus, and the lighting and sound recording system simply must be mentioned. On the other hand, I will also explain how the phases in the development of film are connected to media history in general – not only with the history of other entertainment media, like radio, but also with the development of science and technology and their military applications more generally. This part of the lecture will no longer be concerned only with registering the reactions of writers to film, but rather it appears – according to the work of Thorsten Lorenz – as if cinema, this new ostentatious form of technical display, called the meaning of modern theory itself into question. After all, apart from the old ritualistic meaning of "carnival procession," the word "theory" – the primary word of the Greek philosophers – meant nothing other than "look," "observe," "a feast for the eyes," "a spectacle," or even "pageantry," and it first assumed the meaning of "scholarly teaching" after or through Plato. This is the reason why theorists – who, like Bergson and Sartre, Freud and Benjamin, have become contemporaries of cinema – are confronted with the question of what has become of their spectacles, which exist only as letters in books, under audiovisual conditions. As you have probably already guessed, the technical spectacle did not sit well with the theoretical one.

Like every victory, however, it should not be forgotten that the triumph of film in the twentieth century was temporary. In contrast to the forgetfulness or nostalgia of many film scholars (whereby forgetfulness and nostalgia probably coincide), it should be emphasized that in the eyes of technicians film was already from the start, long before its heyday, seen as only provisional. Mechanical-chemical image recording, mechanical storage, and mechanical playback are out of place in a century that is defined primarily through the conversion of traditional media to electricity. Without Edison's invention of the electric light bulb his film equipment would surely also not have been built, but light bulbs still are not electrical telecommunications. I will therefore show next how electrical telecommunications enabled the transition from silent to sound film – with consequences, incidentally, that extended far beyond technology to the financial structure of the film market. In the second part of these lectures, however, the central issue will be the fully electronic visual medium, and this will require an understanding of the electron tube, which was produced from Edison's light bulb literally behind his back. The fully electronic optical medium, in so far as it has not already been superseded by LCD screens, was based for almost a century on Braun electron tubes: I am speaking of course about television.

As a fully electronic medium, television – if you will pardon this necessary truism – is just as ubiquitous as it is mystifying, and therein lies its much heralded power. I don't know how many of you would be able to operate a television studio or even repair a television set. This technology is so extremely complicated in comparison to film that we are also required to pursue television history in order to learn anything about the *modus operandi* of electronic image-processing from its first tentative steps to today's image standard. These lectures must therefore deal step by step with cable television prior to World War II, black-and-white television after 1945, and finally the three color standards that exist today, without overlook-

24

ing quite unentertaining, namely military devices like radar, which
made civilian television possible. The legal, social, and political
dimensions of the prevailing programming structures are then only
consequences of these technical solutions. I recall an American president who came out of the film industry and who governed with
television interviews.

With the way things are today, however, that is not the end of
the matter, and these lectures will not end there. As an electronic
medium, television is to us (that is, to consumers) only the outward
face of an empire that is already now beginning to strike back. I recall
yet again that president, who not only governed with television interviews but also formulated plans for an optical-electronic future war.
With the introduction of video recorders and video cameras in private
households – and most apparent in the case of computer screens,
which are also in the process of revolutionizing the offices and desks
of lecturers and students – this empire casts out at least a silhouette
of its power. In ten years' time, it is highly unlikely that feature films
will be on celluloid at all, but rather there will only be one form of
standardized opto-electronics, a universal discrete signal processing
that coincides with the universal discrete data-processor known as the
computer. It is my professed goal to conclude this lecture not with
the oldest preserved silent film or with the latest program from RTL,
but rather with computer simulations of optical worlds, no matter
if they enable us to visualize the unreality of mathematical formulas like Benoît Mandelbrot's "apple men" or they hyperrealistically
reconfigure our so-called reality like raytracing or radiosity.[1] And
because such simulations are also the only conceivable future of film
and television for practical and economic reasons, it is important in
this lecture at least to understand the principles according to which
computer programs allow such images to move.

That is all concerning the content of these lectures and the explanation of its title. I have not forgotten the seemingly more obvious
title *Film and Television History*, but rather I have simply avoided
it. Many newly established media institutes at German universities
concentrate or insist on film and television, but this appears to me as a
risky simplification of media technology in general in favor of its most
entertaining and user-friendly effects. In contrast, the title *Optical
Media* should signal a systematic problem and place the general

---

[1] Kittler develops the contrast between raytracing and radiosity in "Computer
Graphics: A Semi-Technical Introduction," trans. Sarah Ogger, *Grey Room* 2 (Winter
2001): 30–45.

25

principles of image storage, transmission, and processing above their various realizations. This general and systematic approach does not result in philosophical abstractions, but rather it reveals underlying structures: when it is made clear first that all technical media either store, transmit, or process signals and second that the computer (in theory since 1936, in practice since the Second World War) is the only medium that combines these three functions – storage, transmission, and processing – fully automatically, it is not surprising that the endpoint of these lectures must be the integration of optical media and the universal discrete machine known as the computer.

This overview has hopefully already shown that the connection of old traditional arts like literature, painting, and theater, on the one hand, with technical media on the other hand should not be mere addition. I am therefore attempting a first, if you will, system design. At Humboldt University, when it was still called by its proper name, Friedrich Wilhelm University, a certain Hegel met every week in lecture room six to put all of the arts that could exist under the conditions of the book monopoly in systematic order according to their form and content, genre and historical progression. And that was indeed a flight of Minerva's owl, which is seen only at sunset: less than ten years after Hegel's death the storage monopoly of books (and thus of philosophical lectures) came to an end with the public presentation of photography. We are therefore confronted today with the insane and probably impossible task of developing a historical and systematic knowledge base for an art and media system under highly technical conditions that would be comparable to the one that Hegel, in an incomparable way, was able to construct under considerably more limited conditions.

This systematic question, and the attempt to resolve it through historical analysis, also comes with a warning: please do not expect a history of directors, stars, studios, and celebrities; in other words, please do not expect a history of film and television, which in the end remains organized around a sequence of titles, just like most literary histories. Apart from the theoretical question as to whether technical media make concepts like the author and the subject obsolete, such a history would be practically useless for me, as I have seen far fewer films than most of you. There are enough special courses in cultural studies that provide film and television histories focused directly on the feature film, which could then also be supplemented by courses in media production.

That is all for my preliminary sketch of how these lectures relate to the subject. As a former Germanist, however, I would also like to address the students of literature and German among you. It should have already become implicitly clear – but to allay any fears I will say it again – that the problems posed by writing and its aesthetics will certainly not be left out. An analysis that examines both the intersections and dividing lines between writing culture and image technology within a historical context is precisely a methodical preparation for the pressing question of what the status of writing or literature can be today.

We know the familiar twentieth-century answers. Under the massive primacy of audiovisual entertainment electronics, which according to many media theorists like Walter J. Ong "brought us into the age of secondary orality" that resembles archaic and therefore allegedly non-literate cultures (Ong, 1991, p. 136), droves of "authors" have taken over the functions of screenplay suppliers or the writers of technical instruction manuals. Others have opted for the opposite approach by claiming that the only serious modern criterion for literature is its structural unfilmability.

While such options require the ability to write and read fluently, we know that audiovisual media have led to a new illiteracy. Like Hölderlin's notion that salvation lies close to danger, however, this new illiteracy may also be the answer to the current status of writing. According to American surveys conducted eight years ago in connection with former president Bush's extremely fruitless educational reform plans, an astonishing percentage of high school graduates are unable to write their own names. Don't ask me what kinds of simulations such students employed to pass their final examinations. The ability to write one's own name was certainly important in the case of Faust, as he signed the pact with the devil, and the ability to sign checks also counts in economic terms; the mere absence of this ability, however, does not explain the horror that accompanied the spread of the new illiteracy. Indeed, the industry that complained about such deficiencies among high school graduates did not seem interested in providing students with a further proficiency in the 26 letters, but rather it devised (typically enough under the overall control of Kodak, the largest photography and film company in the world) an entirely different future for writing. In factories, which in the meantime have all been re-equipped with CAD and Computer Aided Manufacturing, it would be a glaring impossibility to continue employing workers who were unable to master the following literacy skills: to read or draw the flow chart of an electronic circuit, to understand or write a small computer program, to read or even program the graphic display on a computer screen. That is quite literally the

27

job description that the industry itself provided for graduates, and thus also for America itself, in the 1990s.

These lectures, to put it with the requisite innocence, were formulated with a knowledge of this job description. There is every reason to believe, namely, that the educational program they outlined will be implemented throughout at least half of the world. Someone able to master both the old craft of writing and the recently developed technology of digital image-processing would therefore have better career prospects. And for me, that job description of the future serves as a welcome justification for the risky undertaking of dealing not only with conventional film and television but also with the newest technologies like imaging. It appears as if opportunities in the future are expected to be better in the field of video, which will explode through the link to computer technology, than they are for the practically obsolete dream of becoming the last and greatest of all feature film directors. The linking up of a fiber-optic cable network, which will replace the notoriously narrow bandwidth of copper wires, will increase the need for transmittable and processable images just as the need for mythical stories about Hollywood's mythical stories will decrease. It is not without reason that Bill Gates attempted in the last few years to realign his quasi-monopoly on computer operating systems with yet another monopoly on digital images. A Microsoft subsidiary by the name of Corbis travels around all possible museums, archives, and picture collections, generously abstaining from buying any of the stored originals, but receiving for a trifling sum the digital rights for those copies that Corbis itself has scanned (Schmiederer, 1998). And because you can imagine that cities like Florence or even Berlin have more beautiful pictures than Tallahassee or Petaluma, the lion's share of Corbis' loot comes from Europe, which has not yet learned enough about optical media to protect its own digital rights from Microsoft.

More cannot be said about the practical relevance of these lectures. But this also provides a transition to the third point to be dealt with today: the theoretical assumptions and basic concepts I will be working with.

— 1 —

THEORETICAL PRESUPPOSITIONS

The basic concept in the following history and analysis is the concept of the medium in the technical sense, which was developed above all by Marshall McLuhan, whose work was based on the fundamental historical groundwork laid by Harold Adams Innis. This Canadian school, as it was christened by Canadian insider Arthur Kroker (Kroker, 1984), attempted to examine the technical media and the immediacy with which they were let loose on the population of the western hemisphere following the Second World War. According to McLuhan, media are the intersecting points (*Schnittstellen*) or interfaces between technologies, on the one hand, and bodies, on the other. McLuhan went so far as to write that under audiovisual conditions our eyes, ears, hands, etc. no longer belong to the bodies they are associated with at all, let alone to the subjects that figure in philosophical theory as the masters of the aforementioned bodies, but rather to the television companies they are connected to. This connection between technology and physiology, which is not simply dialectical but rather direct, should be recorded and extended. Only McLuhan, who was originally a literary critic, understood more about perception than electronics, and therefore he attempted to think about technology in terms of bodies instead of the other way around. According to the analytical stress model, which had just been discovered at that time, technical prostheses of a sensory organ – in other words, media – were said to have replaced a natural or physiological function, and the biological function itself acted as the subject of the replacement: an eye that is armed with lenses or glasses (a beautiful expression) performs a paradoxical operation, according to McLuhan, as it extends and amputates itself at the same time. In this way, McLuhan is part of a long tradition that can be traced back

29

to Ernst Kapp and Sigmund Freud, who conceived of an apparatus as a prosthesis for bodily organs.

In *Civilization and its Discontents,* Freud above all formulated very drastically how, on the basis of telescopes, microscopes, gramophones, and telephones – as always, Freud does not mention film – so-called modern "[m]an has, as it were, become a kind of a prosthetic god. When he puts on all his auxiliary organs he is truly magnificent," yet he is abject without them because "those organs have not grown on to him" (Freud, 1953–74, XXI, pp. 91–2).

Nothing against this mixture of power and powerlessness, the sublimity and the absurdity of people according to Freud and McLuhan; but their unquestioned assumption that the subject of all media is naturally the human is methodologically tricky. For when the development of a medial subsystem is analyzed in all of its historical breadth, as the history of optical media is being analyzed here, the exact opposite suspicion arises that technical innovations – following the model of military escalations – only refer and answer to each other, and the end result of this proprietary development, which progresses completely independent of individual or even collective bodies of people, is an overwhelming impact on senses and organs in general. McLuhan, who converted to Catholicism long before his international career, hoped to gain something like the redemption of all literature or literary studies from the electronic media of the present and the future. To verify this point, which is cardinal for our context, I cite the following passage:

> Language as the technology of human extension, whose powers of division and separation we know so well, may have been the "Tower of Babel" by which men sought to scale the highest heavens. Today computers hold out the promise of a means of instant translation of any code or language into any other code or language. The computer, in short, promises by technology a Pentecostal condition of universal understanding and unity. The next logical step would seem to be, not to translate, but to bypass languages in favor of a general cosmic consciousness. (McLuhan, 1964, pp. 83–4)

In contrast to such an arch-catholic media cult, which simply confuses the Holy Spirit and Turing's machine, it is hopefully sufficient to point out that the development of all previous technical media, in the field of computers as well as optical technology, was for purposes directly opposed to cosmic harmony – namely, military purposes.

But such a lack of clarity in McLuhan's concept of media should not prevent further work on his fundamental theses. You are presumably familiar with the famous formula that the medium is the message. Without this formula, which virtually prohibits looking for something else behind technically manufactured surfaces, media studies would actually continue to have a subject – just as mysterious fields like theology or World Ice Theory[2] have subjects – but media studies itself would not exist as such in isolation or with any methodological clarity. To determine the concrete subject of media studies one need only connect McLuhan's formula "the medium is the message" – as well as the mock formula he himself came up with in his later years, "the medium is the massage" – with its lesser-known explication that the content of a medium is always another medium. It is therefore obvious that, to take the first example that comes to mind, in the relationship between feature film and television the most popular content of television broadcasts is film, the content of this film is naturally a novel, the content of this novel is naturally a typescript, the content of this typescript, etc., etc., until at some point one returns back to the Babylonian tower of everyday languages.

Taking up McLuhan seems even more advisable because German media studies typically proceeds on entirely different grounds and with entirely different fundamental hypotheses. As Werner Faulstich, one of its leading representatives, repeatedly emphasizes, this media studies sees itself as a direct continuation of the research areas of popular fiction, on the one hand, and the sociology of literature, on the other hand, which both rose to prominence in the 1960s (Faulstich, 1979, p. 15).

Literary scholars who do not forget media would have thus been permitted to remain safely in the native realm of their own intellects; it is doubtful, however, whether such a trivial, content-based approach to media, which are themselves already the message according to McLuhan's contrary thesis, comes near enough to their technical complexity. We would always only be able to grasp the external façade that the electronics industry consciously displays, while the interior of the apparatus would remain concealed beneath its cover, whose instructions permit it to be opened only by an expert. Perhaps

---

[2] A cosmological theory proposed by Hans Hörbiger and Philipp Fauth, which was first published in their 1912 book *Glazial-Kosmogonie* (Glacial Cosmogony). Hörbiger and Fauth claimed that the Milky Way was composed of blocks of ice, and over time these blocks of ice collided and formed planets. They also claimed that the moon was a block of ice, and previous moons collided with the Earth on several occasions, causing the great flood and the destruction of Atlantis. Because it contradicted Albert Einstein's theory of relativity, the National Socialist Party promoted World Ice Theory as an alternative to "Jewish" science.

31

the voluntary self-control of German media studies and its particular focus on trivial or popular content was plausible for so long because on the side of media production itself content and technology fell into separate areas of competence, offices, and organizations. But it is obsolete in the age of the computer, which supersedes this separation on all levels. The only thing that remains is to take the concept of media from there – in a step also beyond McLuhan – to where it is most at home: the field of physics in general and telecommunications in particular. At the beginning of our next meeting, I will attempt to provide you with a systematic introduction to this topic by first of all presenting the basic concepts that Claude Shannon developed in his 1949 mathematical theory of communication – otherwise known as modern information theory (Shannon and Weaver, 1949). What emerges in place of a conglomeration of different media, as German media theorists always still describe it, is a systematic outline, a general connecting thread with which many individual threads could be strung together.

Second, the consequence of employing the media concept of telecommunications is that media studies cannot be limited solely to the study of media that (to be brief and clear) have a public, civilian, peaceful, democratic, and paying audience. For example, in Faulstich's *Critical Keywords in Media Studies* he proposes that closed circuit television systems, like those used for department store security, are of peripheral importance compared to the television, which is more often examined in media studies. That may be statistically true, but it is methodologically unacceptable. For when it can be shown that precisely the civilian and private use of video recorders has arisen from such security systems, it also becomes clear how artificial the dividing line between mass media and high technology is and how much it hinders the analysis of connections. In the end, the categorization of technical media according to their price and their display in department stores only conceals what the late Albert Einstein called the general explosion of information in the present. Einstein was thus strangely (and unforeseeably) in agreement with Heidegger that the explosion of information is more dangerous than all atomic bombs.

When one is methodologically inclined towards a general concept of media and information, though, the problem emerges whether and how some areas can be excluded. For this lecture especially, we are confronted with the problem of acoustic media; although they are not included in the title, they are increasingly networked with optical media. Because the general concept of information is not

32

philosophical but rather technical, which means that it has already ensured its own realization, it is increasingly difficult for telecommunications to be specified and defined through its contents or sensory fields. The development of optical media closely parallels the development of acoustic media, and in some cases they even developed in conjunction with one another. This can be seen in Edison's work on phonography and film and Nipkow's work on telephony and television. Indeed, there would be no television at all if radio technologies had not been developed, which then – after many technical contortions that would never have been necessary for the transmission of voice and music – were also brought to the point where they could be used to transmit images.

After attempting to separate this lecture from sociological and other approaches, what remains are the problems presented by the history of technology itself. In spite of all the metamorphoses of art scholars into engineers, can there be a history of technology at all within the context of cultural studies? In a book about early silent film whose title, *Knowledge is Medium*, is borrowed from McLuhan, Thorsten Lorenz put his finger on the problem: film is simply patent number so-and-so – the plan to build a new device that the brothers Louis and Auguste Lumière submitted in 1895 and that was also awarded by the French government. Every additional word about film beyond this degenerates into cultural or cultural studies gossip. From this, Lorenz decides to take the next logical step and write his obviously cultural studies book not about film but rather about the cultural studies gossip about film.

In our context, however, the suggested practical relevance already excludes such radicality. I will therefore focus on the history of technology and will not exclude comments on patent specifications if only, at the very least, to convey a certain know-how. To a large extent, though, the technical explanations will be oriented towards each of the beginning stages of the development of optical media in order to avoid the difficulties associated with understanding the mathematics. For didactic reasons, it is advisable to present solutions to complicated technical problems at the moment they first emerged, as they are therefore in a condition where they are still easily comprehensible and apperceptible basic circuits, which the inventors themselves must first convert from everyday language into sketches of technical plans, so to speak. In contrast, a television appliance in its contemporary, practically finished form has been through so many development teams and laboratories that it is impossible for anyone to account for all of its individual parts any more.

This emphasis on solutions to early problems runs the risk, as in many histories of film, of falling under the spell of a cult of genius pioneers or inventors and so forgetting the quotidian aspects of the media industry once it is established. But when this developmental history is represented in some detail, as I will attempt, the aura of these individual geniuses dissolves. Not only is genius one percent inspiration and 99 percent perspiration, as Edison once said, but according to McLuhan's law the development of media under highly technical conditions always requires the development of other media and thus the sweat of others as well. One must therefore consider developmental teams, subsequent developments, optimizations and improvements, altered functions of individual devices, and so on; this means, in the end, an entire history of the industry. At this point, though, I immediately recognize my own limits: the history of film and television that I will present does not include the actual history of the industry. I am neither a publicist nor an economist, so I can only deal with the economic and financial conditions of what might perhaps be called the global image trade through hints and references.

In place of the missing history of the industry, which is and remains merely suggested, these lectures will stress two other themes, which follow quite directly from my previous comments on McLuhan. The first concerns the relationship between the history of technology and the body, and the second concerns the relationship between modern technologies and modern warfare.

First, technology and the body: the naked thesis, to place it immediately up front, would read as follows: we knew nothing about our senses until media provided models and metaphors. To make this brief thesis seem plausible, I will give you two extremely opposed historical examples:

a) As alphabetical writing, this new medium of Attic democracy, was standardized on a governmental level in Athens, philosophy also emerged as Socratic dialogue, which the student Plato then put into writing, as we know. Thus, the question was on the table as to which tools philosophers could actually employ. The answer was not the new ionic vowel alphabet, as a media historian like myself would have to answer; rather, the answer was that philosophers philosophized with their souls. All that remained for Socrates and his enthusiastic interlocutors (enthusiastic because they felt flattered) was to explain what the soul itself was. And lo and behold: a definition of the soul was immediately offered by the wax slate, that *tabula rasa* upon which the Greeks etched their notes and correspondence with their slate pencils. Under the guise of a metaphor that was not

34

just a metaphor, therefore, the new media technology that gave rise to the soul was eventually seen as the vanishing point of this newly invented soul.

b) Around 1900, immediately after the development of film, it appears that there was an increase in the number of cases of mountain climbers, alpinists, and possibly also chimney-sweeps who, against the odds, survived almost fatal falls from mountains or rooftops. It may be more likely, though, that the number of cases did not increase, but rather that the number of scientists interested in them did. In any case, a theory immediately began to circulate among physicians like Dr. Moriz Benedict and mystical anthroposophists like Dr. Rudolf Steiner, which even you may have probably heard as a rumor. The theory stated that the so-called experience – a key philosophical concept at that time – of falling (or, according to other observations, also drowning) was allegedly not terrible or frightening at all. Instead, at the moment of imminent death a rapid time-lapse film of an entire former life is projected once again in the mind's eye, although it is unclear to me whether it is supposed to run forwards or backwards. In any case, it is evident: in 1900, the soul suddenly stopped being a memory in the form of wax slates or books, as Plato describes it; rather, it was technically advanced and transformed into a motion picture.

In these lectures, however, the attempt to define the soul or the human being once more will be systematically avoided. As the two examples above quite clearly show, the only thing that can be known about the soul or the human are the technical gadgets with which they have been historically measured at any given time. That excludes the possibility of basing these lectures on the experiences of motion picture audiences and the opinions of television viewers, which most of the work in empirical German media studies continues to be based on (despite all the statistical tricks with which those experiences and opinions are then supposedly transformed into objective data). Fans will therefore not get their money's worth.

Why this disappointment? Because the historical tendency to employ technical media as models or metaphors for imagining the human or the soul, which I have just illustrated, is anything but accidental. Media have become privileged models, according to which our own self-understanding is shaped, precisely because their declared aim is to deceive and circumvent this very self-understanding. To be able to experience a film, as it is so wonderfully called, one must simply not be able to see that 24 individual images appear on the screen every second, images that were possibly filmed under entirely different conditions. This is particularly true of television, as we know that there is a recommended optimal distance between the slipper cinema, on the one hand, and the wing chair, on the other. Eyes that fall short of this distance are no longer able to see shapes and figures, but rather only countless points of light that constitute their electronic existence and above all their non-existence – in the form of moiré patterns or blur.

In other words, technical media are models of the so-called human precisely because they were developed strategically to override the senses. There are actually completely physiological equivalents for the methods of image production employed by film and television, but these equivalents themselves cannot be consciously controlled. The alternating images in film correspond roughly to the blinking of eyelids, which mostly occurs entirely automatically; with some effort, this blinking can be increased to at least half the frequency of film's 24 frames-per-second, which very graphically simulates the stereoscopic effects of film when combined with head movements, but the speed of 24 frames-per-second was intentionally chosen exactly because eyes and eyelids are unable to attain it. In a similar way, the construction of images on television corresponds to the structure of the retina itself, which is like a mosaic of rods and cones; rods enable the perception of movement, while cones enable the perception of color, and together they demonstrate what is called luminance and chrominance on color television. Retinas are themselves seen so rarely, however, that the place where they, and that means all of us, see nothing whatsoever – the blind spot where the optic nerve leaves the eye – was only first discovered by physiological experiments in the seventeenth century.

This implies, conversely, that for technical media, if they impinge upon our senses at all like film or television, it is completely justified to conceive of them as enemies (and without the cultural pessimism that Horkheimer and Adorno's chapter on radio and film in *Dialectic of Enlightenment* made fashionable). For the enemy is, according to Carl Schmitt, the embodiment of our own question. There are media because man is (according to Nietzsche) an animal whose properties are not yet fixed. And precisely this relationship – not a dialectical but rather an exclusionary or adversarial one – ensures that the history of technology is not so ahuman that it would not concern people.

The name for this problem area, which has yet to be negotiated in detail, is standards or norms. Standards determine how media reach our senses. All of the films that can be bought are known to be standardized according to either DIN or ASA. I employ the term "standard" to distinguish those aspects of the regulations that are intentional from the accidental or contingent. Norms, on the other hand, were and are an attempt to cling to natural constants, like the standard meter of the French Revolution, which led medical historian Canguilhem and his follower Foucault to define post-1790 Europe as a culture of norms instead of laws. In this sense, I go one step further and say that after 1880 we find ourselves in an empire of standards (the word culture, as a concept associated with agricultural growth, has to be ruled out). The use of screens for film and panel painting already makes the difference between media standards and artistic styles abundantly visible. This will still be shown in technical positivity, but beforehand I will briefly sketch out the fundamental principles.

The eye sees. Is it seeing a film, a television broadcast, a painting, or a detail from so-called nature that (according to the Greek word) it projects from within itself? This question can only be decided by 1) an observer who sees this eye see, or 2) this eye itself, if and so long as the media standards are still a commercial compromise that reveals deficits, such as black-and-white images, no stereoscopic effects, or missing colors like the American NTSC television system. Like the film director Von Göll in Pynchon's great world war novel correctly said: We are "not yet" in the film (Pynchon, 1973, p. 527).

From the perspective of the year 1945, therefore, Pynchon's fictional director, who is really only a pseudonym for his historical colleagues like Fritz Lang or Lubitsch, promises a standardization that will bring an end to the difference between film and life – like the subtitle of a novel by Arnolt Bronnen – while in the meantime already making some actual advances towards this goal. As you know, this convergence of mediality and reality has been discussed using the term "simulation" at least since Baudrillard. These lectures will have to take up this debate yet again, as the concept of simulation, which refers to the sublation of a separation, allows for the introduction of a sharper distinction between traditional arts and technical media than is customary in everyday language.

In the Greek tradition, there are fairly paradigmatic anecdotes about a competition between two painters, who both claimed to have absolutely fulfilled the allegedly Aristotelian postulate of a μίμησις φύσεως, an imitation of nature. The first painter, who was named Zeuxis, created a painting with remarkably realistic-looking grapes. His competitor was actually able to see that these grapes were painted, but a flock of birds immediately pounced on the painting, thinking that they were indeed real. According to Kant, these two reactions

37

exemplify the entire difference between art and life, disinterested satisfaction and desire. But the matter does not simply end there. It was left up to Zeuxis' competitor, Parrhasios, to take the painting competition to another level. When he presented his work for Zeuxis' assessment, a veil still hung over the painting. Zeuxis wanted to pull the veil away in order to take a better look, but when he attempted to extend his hand towards the veil he realized it was also painted. The first-order simulation was thus able to fool the eyes of animals, while the second-order simulation was also able to fool the eyes of humans.

This anecdote actually shows quite beautifully that art and media are fundamentally about the deception of sensory organs (Lacan, 1981, p. 103), but this seems to be just as beautiful as it is problematic. It implies that people can deceive others about the status of their own creations through the use of manual tools and abilities, such as painting, writing, or composition. "Whoever believes it is possible to lie with words might also believe that it happened here," wrote Gottfried Benn about his early novels. He himself believed it as little as I do. When one sees the remaining Greek panel paintings today, which have admittedly been poorly preserved, the anecdote about the two painters seems very doubtful, as these paintings were obviously done using a palette that included certain colors and simply lacked others. In place of the so-called truth of nature, therefore, these paintings reflect a convention that one must first ignore or overlook in order to fall under the spell of the illusion. In this respect, despite its realistic veneer, painting is not so very different from other arts like music or literature, whose encoding, and this means conventionality, is more readily apparent. The thesis would thus be that traditional arts, which were crafts according to the Greek concept, only produced illusions or fictions, but not simulations like technical media. Everything that was style or code in the arts registered a distinction that is quite the opposite of technical standards.

Artistic styles were certainly ways of acting on the senses of the public, but they were not based on measurements of the abilities *and* inabilities of visual perception like the standard use of alternating images in film; they were based on approximations, conventions, and the pure chance involved in the historical availability of raw materials. Certain artistic effects would not have been possible without oil-based paints, and therefore without petrochemicals and their world wars. If Foucault had been able to write his book about painting as the history of available pigments – as promised in *The Discourse on Language* – we would know more. But it is clear that pigments are just as visible as what they are supposed to show on the canvas. For

38

this reason, European culture up to early modern times was under the control of what Hans Blumenberg called the "postulate of visibility": that which exists also allows itself in principle to be seen (Blumenberg, 1983, pp. 361–75). Plato's concept of theory, which has already been touched upon, even implies that what exists in the highest state of being, namely in the realm of ideas, can itself be seen, although or because it remains absolutely invisible to the naked eye. Technical media and only technical media – according to the thesis of these lectures – destroyed this postulate of visibility. Being, in an eminent sense, allows itself in principle not to be seen today, although or because it allows the visible first to be seen. In this way, the history of optical media is a history of disappearance, which also allows me the freedom to disappear for today.

It is astonishing that the anecdote about the optically deceived birds has returned today in the form of a scientific theory: first, behaviorism has actually established that with female pigeons the ovulation necessary for fertilization occurs not only when they see a cock pigeon, but also when the laboratory deceives them with a two-dimensional dummy. In a second step, French psychoanalyst and structuralist Jacques Lacan then based an entire terminology on the experiment, which has since also made careers among film scholars, particularly in Anglo-Saxon countries. For Lacan, all of the phenomena associated with figure recognition go by the methodological title of the imaginary, and the point is actually that they are just as automatic as they are deceitful. Lacan cites both the pigeon experiment and the ancient painter to support his theory (Lacan, 2002, p. 5), but the example he offers is different: unlike animals, human infants learn from an early age, approximately in the sixth month, to recognize themselves in mirrors. The point of this early childhood figure recognition, however, is that it is also simultaneously a misrecognition – simply because the apparently superior sensory capabilities of human infants as compared to those of baby animals, who see adversaries in the mirror rather than themselves, are inversely proportional to or compensate for their delayed motor skills. It is precisely because they are not yet able to walk and their immature central nervous systems have not yet registered the unity of their own bodies that they project a closed, visually perfect identity onto the mirror image. The tremendous joy they express upon recognizing themselves in the mirror conceals the reality that their bodies are still physically uncoordinated. According to Lacan, this is how the ego itself emerges from the imaginary. And the fact that Lacan found proof for his theories in a scientific experimental film that demonstrated this process of self-(mis)recognition in the mirror also clearly shows how the mirror stage and the imaginary are related to film. I will return to this complex in my discussion of early German silent films, which were full of mirrors and *doppelgängers*.

For the moment, it is more important to emphasize the idea that the imaginary represents only one of the three methodological categories of the structuralist theory. According to Lacan, the dimension of code, which I have just illustrated through artistic styles and aesthetic rules, appears under the title of the symbolic, which turns out to be essentially at home in the code of everyday language.

The third and final category is called the "real," but please do not confuse this category with common so-called reality. *Le réel* refers only to that which has neither a figure, like the imaginary, nor a syntax, like the symbolic. In other words, combinational systems and processes of visual perception cannot access the real, but – and this is one of the leitmotifs of these lectures – this is precisely why it can only be stored and processed by technical media. The present can be distinguished from every earlier period by the fact that we live at a time when, with the help of Mandelbrot's fractals, clouds can be calculated in their full randomness and then be made to appear on computer screens as calculated, unfilmed images. Practically speaking, however, this means that we must employ a considerable part of film theory – which usually goes by the name of film semiotics – in order to clarify the radically new ways in which optical media handle the symbolic. This concerns, more concretely, techniques of montage and editing, and thus everything that has been regarded as specific media aesthetics since the time of Walter Benjamin. Above all, it must be made clear how media, in contrast to all of the arts, can nevertheless include the impossible real in their manipulations, techniques, or processes, and thus treat the pure chance of a filmed object or a television camera setting as if it had the same structure as the manipulable codes in the arts. To shed some light on this possibly vague suggestion, I will conclude these comments on media technologies and the body with a quotation from Rudolf Arnheim's film theory. In an essay on the systematics of early cinematographic inventions, Arnheim claims that "since we have known photography" there has been a new and "more ambitious demand placed on the image": "It is not only supposed to resemble the object [as in all representative arts], but it is also supposed to guarantee this resemblance by being the product of this object itself, i.e. by being mechanically produced by it – in the same way as the illuminated objects in reality mechanically imprint their image onto the photographic layer" (Arnheim,

40

1977, p. 27). This passage hopefully shows what a manipulation of the real can be in contrast to all figures and cultural codes. And if the body belongs to the real, as Lacan argues, then this introduction to optical media and the body is where it should be.

In terms of methodology, it only remains to be noted that I am employing Lacan's terms as a useful set of conceptual tools, not as immutable truth – for the simple reason that over the course of the semester we must ask whether the basic concepts of current theories are absolutely independent and thus true frames of reference or rather a direct result of the media explosion of our own epoch. Lacan's notion of the symbolic as a syntax purified of all semantics, meaning, degrees of figuration, and thus also every conceivability could in the end coincide with the concept of information in telecommunications.

The question still remains as to where the untraditional concept of information itself – the basis and goal of all technical media – originally comes from. To get to this, as well as the relationship between media and war, I will stay with the example of photography and quote an extremely early passage from 1859, in which (as far as I can see) something like media-technical information appears for the first time. Oliver Wendell Holmes, Sr., the first real theorist of photography, wrote at that time:

> Form is henceforth divorced from matter. In fact, matter as a visible object is of no great use any longer, except as the mould on which form is shaped. Give us a few negatives of a thing worth seeing, taken from different points of view, and that is all we want of it. Pull it down or burn it up, if you please. (Holmes, 1859, p. 747)

According to Holmes, therefore, modern information conceals itself under the ancient philosophical concept of form: the possibility of storing, transmitting, and finally processing data without matter and also without the loss of accuracy that was unavoidable in artistic reproductions. The point of his example, however, is only that chemically pure information becomes a correlate of chemically pure destruction. What Holmes is describing already sketches out the path to the bomb over Hiroshima, which, according to the similar insights of Thomas Pynchon and Paul Virilio, represents both a photographic flash and an annihilation, or that Black Forest mine station where the plans and photographs of all of our monuments have been stored in bomb-proof shelters by the federal government of Germany.

In other words, the concept of information itself has a military, strategic component. It is no accident that the age of media technologies is at the same time also the age of technical warfare. French archi-tecture and military theorist Paul Virilio has made this point quite clearly, especially in the case of optical media. In Germany, however, he is overlooked by most media theorists – with the exception of Heide Schlüpmann and her lucid discussion of silent film and World War I. These lectures must and will therefore satisfy a pent-up demand to catch up with his work.

Virilio's argument, above all in his book about war and cinema, follows two separate tracks: the first concerns everything that optical media produce that can be considered imaginary, in the sense that I have just defined, such as all the means of fascination, blinding, dis-guise, or – to use a term from this media-technical century – optical illusions in general. And because Virilio defines war first of all as basically a game of hide-and-seek between two enemies, he is able to show how media effects are linked to military stratagems through optical illusions. This appears to be a rather simple model to explain the present global image trade and image war. For this reason, I prefer to follow Virilio's second line of argumentation, which espe-cially concerns optical media. In contrast to sound waves, which are known to cover a distance of approximately 330 meters per second at normal temperatures (and completely ignoring the speed of letters or orders sent by mail or pony express), the speed of light waves or light particles is Einstein's constant c, which cannot be surpassed by any other speed. Accordingly, Virilio's second argument is that the strategic interest in faster information – the supervising and direct-ing of one's own troops, the monitoring and surveillance of enemy troops, and above all the supervising and directing of one's own response to enemy actions, which should be as immediate as pos-sible – crucially accelerated the explosive rise of optical media over the last hundred years.

It seems necessary to absorb this point and trace it through film and television to the digital future of image technology. I will thus attempt to pass on the factual evidence Virilio has laid out, which in other contexts has been simply ignored, and on the basis of this evidence I will attempt to demonstrate the plausibility of his often radical theories, such as his claim that between the wars popular cinema was (to use Eisenhower's famous phrase) a military-industrial complex.

This naturally implies, as has already been emphasized, that the list of technologies to be addressed does not end with popular films and television programs, but rather the category of optical technol-ogy also encompasses such cryptic things as radar or night vision devices. In these times, when a wall separating Germany has fallen, perhaps it is also possible to conceive how relative every distinction between civil and strategic image technologies has gradually become: apart from the eastern European delay in informatics and computer-controlled production, which Gorbachev himself admitted and which he described as a motive for opening his country to the West, this wall fell as a result of a constant 25-year bombardment of television broadcasts.

And such events, which are triggered by technical media, possibly represent the conclusion of more than just a chapter in postwar European history. Perhaps telecommunications brings history itself, which was always a metaphor for the possibility of written inscription, to a point beyond which it is no longer history in the traditional sense. In any case, it is worth reconstructing the history of film and television within this context. After all, events that exist as nothing but documentary films or television recordings (Kennedy's murder in Dallas, the attack on Reagan in New York) continue to multiply. Such events can no longer be traced back to other, historically correct (that is, written) sources, just as it is also impossible to magnify the corresponding film documents any further without ending up in the pure grain of the celluloid and therefore in a white noise where there is nothing more to recognize (as Antonioni proves in *Blow Up*). It could thus be said that whereas history has handed down to us the opposition between writing – a manual art – and the ocean of undocumented events that remain inaccessible, this is precisely where the new opposition of the media age between technical information and white noise, the symbolic and the real, emerges.

Now that the concepts of information and noise have at least been introduced, I can finally conclude this methodological introduction, as promised or threatened, with a brief sketch of Shannon's technical model of communication and information.

Claude Elwood Shannon, a leading mathematician and engineer in the research laboratory of AT&T, which remains at present the largest telephone company in the world, outlined this model in 1948 in a work with the modest yet equally ambitious title *The Mathematical Theory of Communication*. After the Second World War brought about a surge of innovations in all fields of telecommunications, particularly in television and radar, it became historically necessary no longer to produce theories about individual media, as everyone had done for film from Hugo Münsterberg to Walter Benjamin, but rather to pose the simple and common question of what media technologies in general do; what are their functions and constituent elements

43

that enable information even to occur? Shannon was able to answer this question thanks to mathematics and its elegance. Although this mathematical aspect must be toned down for it to come in useful here, for our purposes it offers the advantage of introducing clearly delineated concepts that make it possible for the first time to compare the performance and limits of individual media, like film and television, with each other. Once the general functions and elements are known, they can be found at the most varied degrees of technical complexity, from the simple, old-fashioned book to the newest computer screen.

There are five interconnected elements in Shannon's general model of a communication system: first, a data source that generates a message; second, one or more senders that translate this message into signals according to the rules of a prearranged code so that the system is able to transmit them; third, a channel that actually conveys the transmission (with a considerable or slight loss of information); fourth, one or more receivers that treat the signal in the opposite or inverse way to the senders, if possible, and reconstruct or decode the message from the received flow of signals; fifth and last, one or more data sinks to whom, Shannon writes, the message is addressed. According to the mathematical theory of communication, it is completely unimportant what kinds of entities serve as data sources that transmit a message and data sinks that receive a message, such as humans or gods or technical devices. In contrast to traditional philosophy and literary studies, Shannon's model does not ask about the being for whom the message connotes or denotes meaning, but rather it ignores connotation and denotation altogether in order to clarify the internal mechanism of communication instead. At first glance this appears to be a loss, but it was precisely its independence with regard to any sense or context that allowed technical communication to be emancipated from everyday languages, which are necessarily contextual, and that led to its global victory. When Shannon explicitly says that we have no need for a communications system for eternal truths, whether of a mathematical or even, I would add, religious nature because such truths must be continuously reproducible at different times and places without technical transmission, it becomes abundantly clear how the essence of media diverges from our everyday concept of faith. Let us therefore forget humans, language, and sense in order to move on to the particulars of Shannon's five elements and functions instead.

Because it is conceived without reference to any semantics, the message can be of an arbitrary type: a sequence of letters as in books

44

or telegraph systems, a single quantity that changes over time like the vibrations of voices or music on the radio or on records (if we disregard the two variables of stereophony), or, in an extremely complex case like color television, it can also be an entire conglomeration in multiple dimensions of both space and time. For a single color image to be seen, the two spatial dimensions of a red value, a blue value, a green value, and a brightness value must be transmitted at the same time as the temporal dimension of sound.

The sender, the second link in the chain, predictably has the function of serving as the interface between the aforementioned message and the technical system; it must therefore find a happy medium or compromise between the complexity of the message and the capacity of the channel. In principle, there are two possible solutions: in the first case, the signal generated by the sender corresponds proportionally to the message, which means that it follows all of its changes in space and/or time. This is called analog communication, as in the case of gramophone, microphone, radio or even photography, and while it is more familiar it is also unfortunately more difficult mathematically. In the second case, the message is broken down into its pure constituent elements prior to transmission in order for it to fit the capacity of the channel, which is in principle always physically limited. These elements are entirely of the same type, such as letters in the case of a spoken message or numbers in the case of computer technology or the individual pixels of a monitor. Because these elements can only assume certain values – there are, for example, far fewer Latin letters than the number of possible sounds produced by the larynx and mouth – they cannot match all of the variations, intricacies, and details of the message. Communication systems that employ such mathematically and technically verifiable signals are called discrete or – following the model of the finger of a hand – digital.

And the entire difference between film and television studies will amount to the clarification of how the transition of a largely analog medium like film to the digital television screen changes or revolutionizes visual perception.

Third, the channel is equipped for the technical bridging of space in the case of transmission media or of time in the case of storage media, and it can consist of material, like telephone wires or fiber-optic cables, or it can simply be a vacuum through which electromagnetic waves propagate, like radio or television. As a physical medium, in any case, every channel also generates interference or noise, which is the conceptual opposite of information. When a television is set on a frequency between the regular channels, this noise appears to our

45

sensory organs (which are otherwise blind to noise) like snow made of points of pure light that correspond to some accidental event like spark plugs or distant galaxies. It is impossible to determine whether the noise represents a single ongoing accidental process or the sum of an endless number of such processes. In any case, for all media the technical specifications must aim to reduce the level of noise in the channel – eliminating noise altogether is impossible – and increase the level of signal. And Shannon's theoretically crucial computational result was that this is primarily possible by cleverly coding messages and repeating them until they are received with the desired level of accuracy.

Fourth and last, the task of the receiver in a communication system is to decode the technically encoded signal and thus reconstruct as far as possible and feasible the message submitted from the sender. In the case of a book, this amounts to the simple act of reading. In the case of technically complex media like television, on the other hand, an electronic signal that is not perceived by any sensory organ must first be transformed back into a form that to some degree accommodates the physiology of our eyes. In the case of digital media, like electronic image-processing, this transformation requires a digital-to-analog converter to allow for human sensory organs. What one sees in the end is therefore only the outer onion skin of an entire series of conjuring tricks that must first be invented, calculated, and optimized, and Shannon drew up formulas for precisely these calculations, which can be applied to absolutely all technical media in general. If you have noticed, like movie fans for instance, that in my lecture about the five functions of communication the seemingly fundamental and necessary function of storage does not appear in Shannon's work at all, I can only respond in two ways: first, the function of storage is concealed but also thoroughly explained by the mathematics of code optimization, which I mentioned only fleetingly, and second, it is probably an indication of our own situation if all media, as in Shannon's work, are defined as transmission rather than simply storage media. While the purpose of a Christian festival like Easter is to be ritually repeatedly every year simply because it is supposed to store and transmit a fixed and well-known message, namely the gospel (good news), no one is particularly pleased about repeated broadcasts on the television. Shannon's technique of measuring information mathematically was specifically developed to distinguish and determine the newness or improbability of a message compared to the mass of repetitions that are necessarily implied in every code.

— 2 —

TECHNOLOGIES OF THE FINE ARTS

2.1 *Camera Obscura* and Linear Perspective

2.1.1 *Prehistory*

By panning from Christmas to prime time television, from the Christian message to technical signals, I have already arrived deep in the prehistory of optical media. To express it in one sentence: today images are transmissible; however, over the course of history images, at least in principle, could only be stored. An image had its place: first in the temple, then in the church, and finally (to Heidegger's dismay) in the museum. And because this place – according to Benjamin's theory of the aura – was far away, perhaps even "the unique phenomenon of a distance" (Benjamin, 1969, p. 222), there was at best the possibility of a museum visit or an image trade and at worst the possibility of an image theft. Writing, on the other hand, served not only as a storage medium for everyday spoken language, but also (I admit) as a very slow broadcast medium after the practice of inscribing on walls or monuments was superseded by the use of papyrus and parchment. Books can be sold, sent, and given away. Writing was therefore not merely literature but always mail as well. And the evidence supports the assumption of Harold A. Innis, McLuhan's predecessor in media studies, that it was the portability and transmissibility of scrolls that brought the two nomadic tribes, first the Jews and later the Arabs, to replace the worship of extremely heavy images of god with a god-given or even god-written book (Innis, 1950). The Bible and the Koran were only able to begin their victory march against all the temple statues and idols of the Near East and Europe because they were mobile relics. Because writing combines storage and transmission in a unique way, its monopoly held sway until media made letters and numbers, images and sounds technically mobile. Instead of reeling off the prehistory of film and television as a report about the large and small steps made by various inventors, therefore, I will frequently represent it with regard to literature, whose monopoly the new media first had to defeat.

Most accounts of this prehistory begin with a discussion of cave pictures from the Stone Age or Egyptian funerary inscriptions, which supposedly were an attempt to capture sequences of movements from so-called nature in an image or a series of images by artisanal means. This is the approach taken by Friedrich von Zglinicki, for example, in order to reel off the content of his highly commendable book *Der Weg des Films* (The Way of Film) like a film.

Or more precisely: like a Hollywood movie. For Zglinicki begins with the thesis that humanity as such – this monstrous collective singular – always harbored the old dream of making images move and thereby making them even more like reality (Zglinicki, 1979, p. 12), and the invention of the first technically functional film in 1895 thus represented the happy ending of this dream. In contrast to such a history of motion pictures, which must bend all possible facts, one could derive the exact opposite lesson from film technology itself: rather than tracking the same continually moving image over the course of millennia, like Zglinicki, one could conceive of these millennia themselves as film cuts. I therefore try to emphasize instead the caesuras or breaks in perception and artistic practices that were necessary in order to reach the threshold of moving images.

The problem of moving images is once again the problem of image transmission, simply because movement as such cannot be stored without media and transmission itself is a type of movement. What remained in the absence of technology was the use of the very short and unreliable channels, which the physical laws of nature are able to provide. All the myths based on the shadow and the mirror revolve around the problem of image transmission. The ancient gods found the answer to this problem easy, because they themselves were already statues in the temples: their reflections, doubles, and image transformations, which mostly followed erotic goals much like popular movies, fill an entire mythology, instead of which I am only able to recommend Pierre Klossowski's *Diana at Her Bath*.

It is more difficult for mortals to manipulate moving images, or even to see them clearly as images at all. I remind you of the story of the young Narcissus from Ovid's *Metamorphosis* and the allegory of the cave that Plato devised in his *Republic* and that film theorists since 1920 – from Paul Valéry to Luce Irigaray – want to read as the model of all films. But Narcissus fell in love with his own reflection in the surface of a pond precisely because this "simulacrum" made the same fleeting gestures as he did himself. And the simulacrum in the allegory of the cave reproduced everyday objects, whose clay or wood imitations were carried past a single light source by a pair of puppeteers behind the backs of the bound cave occupants. On the wall before their eyes, only fleeting shadows come and go, which do not come to rest anywhere. To count the allegory of the cave as a precursor of film is thus absurd. Because moving images could not be stored in his own time, Plato equated the immortal and therefore self-storing soul with a wax writing slate, the medium of his own philosophy.

This writing monopoly saves us the trouble of treating simple technical realizations as optical media, such as Javanese shadow puppet theater or the mirror effects of the ancient *deus ex machina*, which was supposedly made to appear at cultic festivals through a mechanism invented by Heron of Alexandria. Instead, we can move directly to the first solutions to the problem of how a transmitted image could also be made to store itself. As the great physicist Du Bois-Reymond discovered in 1850, from the middle of the fifteenth century onwards scientists and artists have been investigating the question of how "to make nature depict itself, so to speak" (Busch, 1995, p. 90).

What is meant by the self-depiction of nature is so-called "linear perspective," a technique employed in painting since approximately 1420, which ensured that all of the lines, corners, and proportions in an image appear exactly the same as the image they reproduce on the retina. Painting thus became the engineering of illusions, because a more or less explicit geometry stands behind every painted image. Between the Renaissance and Impressionism, this geometry absolutely dominated painting as an artistic style in the aforementioned sense of the word, and since the arrival of photography it has also been incorporated into media technologies as a technical standard.

The question remains as to why this geometry was not always dominant, but rather first emerged at a well-defined time. In Egyptian painting, there existed only a radical joining of frontal and side views, as we know, but Greek pottery painting was also unable to create spaces whose lines all ended at a vanishing point on the horizon. A few perspective effects appeared only in wall-paintings excavated in Pompeii, typically in the arts and crafts ambience of the bedrooms and mysterious cults of late antiquity, yet they obviously still belong to a thoroughly constructed geometry.

2.1.1.1 Greeks and Arabs

There are good reasons for this. In addition to countless other sciences, classical Greece also founded a science of optics, which at least managed to establish the law of reflection, if not also the law of refraction. At the latest since the time of Euclid, who in addition to his famous *Elements of Mathematics* also wrote about optics and the path of light (Edgerton, 1975, p. 68), it was clear to the Greeks that rays of light travel in straight lines. But with the notable exception of the materialistic and therefore atomistic school of philosophy, the ruling doctrine amounted to the foundation of all optical laws on visual rays, which did not lead from the light source to the eye (as in today's understanding), but rather in the exact opposite direction from the eye to the light source. The eye itself thus functioned like a spotlight, whose beams encountered or edited the visible world and then registered this information in the mind. Goethe had good reason, in his great enmity towards Newton's modern optics, for writing this Greek, all-too-Greek verse: "Were our eyes not like the sun, they could never see it." For a discussion of the insurmountable barriers to research that this theory put in place, please consult the work of Gérard Simon (Simon, 1988).

What matters here is only that this ancient theory of active visual rays effectively excluded or prevented any conjecture about the self-depiction of nature. In a closed and finite world, which the Greeks honored with the name cosmos, meaning a well-ordered sphere, these rays could easily reach everything, even the stars that populated the inner surface of the sphere itself, and at the same time the speed of light was also considered infinitely great. Linear perspective, on the other hand, was based on the implicit (and later entirely explicit) assumption of an infinite universe, which corresponded to an infinitely distant vanishing point in every single perspective painting; these paintings thus functioned as miniature models of the infinite universe itself. In a lovely book entitled *Signifying Nothing*, Brian Rotman attempted to grasp this infinity as the intrinsic value of modern Europe from the introduction of zero: first, the vanishing point of linear perspective; second, the zero from the numbers imported from India and Arabia; third and last, the money of modern financial systems – they all supposedly stand for that extremely tricky mathematical function that divides one by infinity (Rotman, 1987). But as you already know: what is forbidden in theory can have explosive consequences in practice. Europe, with all its states, colonies, and sciences, is possibly only the effect of a miscalculation.

Rotman's claim that the cause of these explosive results can be traced back to an Arabic import is no less valid for linear perspective than it is for the decimal place system of modern mathematics. It appears, namely, that a passing comment by Aristotle led Arabic mathematicians like al-Kindi or Alhazen to construct the first workable models of a *camera obscura*, which were thus also the first models of linear perspective. In his so-called *Problemata* (so-called because it was apocryphal), which was hardly more than a collection of notes concerning unsolved questions, Aristotle, who wrote about everything that was knowable *circa* 350 BC, noted not only his still momentous thesis about genius and insanity, but also a small experiment involving a solar eclipse, when it appears, from our earthly perspective at least, that a full moon moves directly before the sun. In antiquity, however, the sun was not defined merely by the fact that it makes everything visible except itself, as looking at the sun leads to blindness. This is precisely what Leonardo means when he says, in the passage I cited at the very beginning of these lectures, that the sun never sees a shadow. Despite or because of this, Greek mathematics had precisely begun, to the astonishment of oriental despots, to be able to predict future solar and lunar eclipses. They were thus able to see exactly what was forbidden to the mortal eye. Aristotle described the simple trick of avoiding the danger of blindness using optical filters. Instead of observing the partially covered sun directly, he recommended observing the entire scene on the rear wall of a room whose front wall contains a small hole.

Aristotle had thus already explained the principle underlying every *camera obscura*, but he only applied it to the special case of the sun, a light source superior to all others. His Arabic translators or successors were the first to investigate the aforementioned hole under empirical, and therefore terrestrial, conditions. In place of the divine sun they employed a simple wax candle, which sent its own light through the medium of that small hole and reproduced an image of itself within the chamber. A *camera obscura* for any light source did not actually exist in the world, but only on paper, yet this paper supposedly reached Europe through an Arabic mediator. In stark contrast to Greek mathematics, Arabic mathematics generally investigated all the possible ways in which obliquely placed light sources encounter the resistance posed by opaque objects and then project the shadow of those objects on vertical walls. (The Greeks limited their curiosity to that quasi-horizontal surface known as the ground in order to be able to determine the time of day from the length of a sundial's gnomon and its shadow.) In the realm of fairy tales,

51

such research produced Harun al-Rashid's trigonometry, which was nothing more and nothing less than a new type of mathematics. Sine and cosine, tangents and arc tangents are all – the things as well as the words – Arabic innovations. Before approximately 1450, when Europe applied all of these trigonometrical functions to very practical purposes, namely the military or colonial navigation of ships, they were first designed with the theoretical purpose of investigating the effects of light rays on flat surfaces. In the *camera obscura* model, for example, the tangent corresponded exactly to the projected length of the reflection of an object standing at angle x in relation to the plane of the camera (provided that all circles are idealized as a unit circle following Leonhard Euler).

In any case, neither Arabic mathematicians nor their European students – the most significant being the Nuremberg patrician Regiomontanus – had anything other than simple empirical methods of conveying such trigonometrical functions. In modern language, such functions are transcendent: they disdain all simple calculations. Sine and cosine, tangent and cotangent were thus available in endless tables, which consisted moreover of huge integers prior to Simon Stevin's invention of decimal numbers. Because he was unable to write up numbers like 0.7071 (the sine of 45 degrees), for example, Regiomontanus multiplied all sine values by a factor of ten million (see Braunmühl, 1900, p. 120). But such monstrous tables of monstrous numbers were virtually unusable by artists, and consequently the history of linear perspective, at least in its first centuries, is certainly not the history of mathematics. I will later come back to the question of when and through whom this changed.

2.1.2 Implementation

But even this mathematical weakness of early trigonometry was able to help the *camera obscura* achieve tremendous success during the Renaissance. As a device that calculated trigonometrical functions completely automatically, simply because it focused light into a single bundle of straight lines and then allowed them to follow their course, the *camera obscura* made the revolutionary concept of a perfect perspective painting possible. Devices, then as now, relieve humans of the need to calculate. However, perspective painting, which was unknown to the Egyptians and the Greeks, was only made possible by going one final step beyond Aristotle and Arabic optics: the *camera obscura* did not simply reproduce light, whether it was the great heavenly sun of Aristotle or the small earthly candle flame used by

52

the Arabs, but rather it also visibly projected objects illuminated by the light. A problem thus emerged that was not completely solved until the invention of optical lenses at the beginning of the seventeenth century.

The *camera obscura*, to use Shannon's rigorous terms, works as a noise filter: the small hole through which the rays emanating from all light sources are forced – directly illuminating lights as well as indirectly illuminated objects – also blocks the scattered lights that are otherwise omnipresent and thus makes the reflection sharp. Otherwise, the image in the *camera obscura* would appear as impressionistic as when the summer sun illuminates the woods. The gaps between the leaves of every deciduous tree function like countless out-of-focus *camera obscuras*, and the end result is that a patchwork carpet of completely contradictory projections emerges on the forest soil – an effect, as I have said, that interested Manet more than the artist-engineers of the European Renaissance. In the interests of their royal and religious patrons, these artists did not want and were not supposed to paint bourgeois picnic breakfasts, but rather a geometrically exact view of the world in general and their own architecture in particular. They thus ran into a problem that absolutely concurred with Shannon's claim that the filtering of a signal always simultaneously also implies the weakening of a signal. The smaller the hole in the *camera obscura*, the sharper but also darker the image becomes; the bigger the hole, on the other hand, the brighter but also the more blurred the image becomes. It is therefore no wonder that the first descriptions of a functional *camera obscura* came from Italy, the western European country with the brightest sunlight: Leonardo supplied the first model around 1500, and Giambattista della Porta, the universal scientist and magician, supplied a more detailed model around 1560. Porta simply suggested darkening the window of a room that opens out onto the sunny side of the street, yet leaving a hole that casts ghostly images onto the opposite facing wall of passers-by and domestic animals floating on their heads. Plato's allegory of the cave was thus implemented.

A gap of 200 years separates Leonardo and Porta from the late medieval references to the *camera obscura* by Roger Bacon (who will also come up in connection with the invention of gunpowder), yet it is precisely in this gap that the invention of linear perspective occurs. Contrary to all of my stories, therefore, the invention of linear perspective would hardly seem to be based on the technology of functioning *camera obscuras*. This gap or hole in the historical record, which fundamentally involves the invention of a hole, can only be filled with historical speculation, which at least has the advantage that it concerns a unique and actually existing hole.

2.1.2.1 Brunelleschi

The history I will now tell concerns one of those great artist-engineers produced by the Italian Renaissance: Filippo Brunelleschi. In contrast to later artists, who remained only artists, artist-engineers were people like Brunelleschi, his younger friend Alberti, or even Leonardo, who were not satisfied with merely producing image after image, but rather for the first time ever they established the artistic and technical standard according to which countless images of an epochal style became possible and feasible. The word "image" here should not be misunderstood to refer solely to the strange two-dimensional pictures on the walls of churches, palaces, and later museums, but rather also to such abstract yet brutally effective things as fortresses or church domes.

Filippo Brunelleschi was born in Florence in 1377. At that time, it was mandatory for novice craftsmen to serve an apprenticeship, just as it is today, and Brunelleschi served his under a goldsmith. In 1401, while presumably still an apprentice or journeyman, Brunelleschi participated in a competition sponsored by the *Signoria*. The Bapistry, the baptism chapel dedicated to John the Baptist that faces the Cathedral in Florence, was to be ornamented with new bronze doors. Although his design featuring the sacrifice of Isaac (which still exists today) was unsuccessful, Brunelleschi's loss was Europe's gain. For instead of maintaining a sole focus on reliefs or art more generally, as his medieval predecessors did, Brunelleschi went on to study mathematics and architecture. Like all of the fortresses that Brunelleschi built as head engineer, the technically incredible dome that adorns the Santa Maria del Fiore, otherwise known as Florence Cathedral, was based on precise mathematics. He died in 1446, barely a year before the impoverished Mainz patrician Johann Gensfleisch zum Gutenberg printed his (presumably) first calendar with movable type. I will soon come back to this coincidence.

But first I want to discuss a small and, more importantly, missing work by Brunelleschi, which at first glance appears trivial in comparison to his domes and fortresses. The fact that we even know about this missing image, which was presumably made in 1425 (Edgerton, 1991, p. 88), is solely thanks to the significant fact that simple crafts-men like Brunelleschi – in total contrast to the anonymity of the European Middle Ages – received the honor of having a biographer in 1450. A description of the work can thus be found in Antonino di Tuccio Manetti's account of Brunelleschi's life:

About this matter of perspective, the first thing in which he displayed it was a small panel about half a *braccio* square on which he made a picture showing the exterior of the church of S. Giovanni in Florence. And he depicted in it all that could be seen in a single view; to paint it he took up a position about three *braccia* inside the middle door of S. Maria del Fiore. The work was done with such care and accuracy and the colors of the black and white marble were so faithfully reproduced that no miniaturist ever excelled him. In the picture he included everything that the eye could take in, from the Misericordia as far as the corner and the Canto de' Pecori on one side to the column commemorating the miracles of St. Zenobius as far as the Canto alla Paglia and all that could be seen beyond it on the other. And for what he had to show of the sky, that is, where the walls in the painting stand out against the open air, he used burnished silver so that the actual air and sky would be reflected in it and the clouds also, which were thus seen moving on the silver when the wind blew. Now, the painter had to select a single point from which his picture was to be viewed, a point precisely determined as regards height and depth, sideways extension and distance, in order to obviate any distortion in looking at it (because a change in the observer's position would change what his eye saw). Brunelleschi therefore made a hole in the panel on which the picture was painted; and this hole was in fact exactly at the spot on the painting where [in reality] the eye would strike on the church of S. Giovanni if one stood inside the middle door of S. Maria del Fiore, in the place where Brunelleschi had stood in order to paint the picture. On the picture side of the panel the hole was as small as a bean, but on the back it was enlarged [through the thickness of the panel] in a conical shape, like a woman's straw hat, to the diameter of a ducat or slightly more [i.e. 2.3 cm]. Now, Brunelleschi's intention was that the viewer, holding the panel close to his eye in one hand, should [turn the picture away from himself and] look [through the hole] from the back, where the hole was wider. In the other hand he should hold a flat mirror directly opposite the painting in such a manner as to see the painting reflected in it. The distance between the mirror and the other hand [holding the panel] was such that, counting small *braccia* for real *braccia* [i.e. measured in the same scale as that which obtained between the painting and the real thing], it was exactly equivalent to the distance between the church of S. Giovanni and the place where Brunelleschi was assumed to be standing when he painted it. Looking at it with all the circumstances exactly as described above – the burnished silver, the representation of the piazza, the precise point of observation – it seemed as though one were seeing [not a painting but] the real building. And I have had it in my hand and looked at it many times in my days and can testify to it. (quoted in Battisti, 1981, pp. 102–3)

This story emphasizes, like no other, what was revolutionary about the new world view called linear perspective. Brunelleschi shattered or literally bored through the entirety or the imaginary nature of a panel painting in order to reveal something even more imaginary. His image of the Florence Baptistry, whose bronze doors he himself wanted to design, proves to all his disbelieving colleagues and contemporaries that perspective vision really and truly always already takes place in the eyes. Otherwise, the eyes would not be so fooled by their own simulation, as Manetti showed Brunelleschi's contemporaries. The fact that such a literal *demonstratio ad oculos* must have been necessary at that time, yet unnecessary today, in the age of fish-eye cameras or satellite images, already says something about Brunelleschi's experiment.

But there is still plenty left to discuss: first, in terms of media history, which images were abolished by Brunelleschi's hole; and second, how could such a perfectly deceptive image have been achieved in 1425?

To begin with the first question, I must go back a little further. As we know, 90 percent of all the images and stone buildings commissioned in Europe in the centuries prior to 1425 were designed to serve the only true Christian faith. This faith happily adopted the Greek Catholic concept of visual rays, which make the world visible to begin with. But this eye, which can still be seen today on any dollar bill, does not belong to any human, but rather to God himself. According to Abbot Suger of St. Denis, the glass windows of the Christian church put precisely this divine visual ray in the picture. God thus presented himself in art – and from his own perspective rather than the distorted perspective from which earthly beings could look at him. For this reason, the icons of the Byzantine Empire – the prime example of the nexus between art and worship according to Hans Belting (Belting, 1994) – principally showed God in front of a gold background that truly implemented his radiance. And, as Samuel Edgerton wonderfully demonstrated, it is precisely this golden background that turned into the first proto-perspective medium in Western Europe. Christian philosophers like Roger Bacon, who has already been mentioned in the context of the *camera obscura*, represented the sacred being as an emanation or radiation of small golden bodies, or *corpuscula*, that travel from heaven into the eyes of humans and thus also into the eyes of those who look at the image. Bacon even employs the Latin word *perspectiva* in order to use a visual metaphor to explain how God's grace spreads throughout the world (Edgerton, 1991, p. 44). In paintings created by devout Italian monks, this journey can also be seen in the form of small golden rings that become detached from the body of the sacred being.

So much for the background information needed to better understand Brunelleschi's revolution. In his book *Belichtete Welt: Eine Wahrnehmungsgeschichte der Fotografie* (Exposed World: A Perceptual History of Photography), Bernd Busch writes: "Brunelleschi's experimental design was revolutionary because it established the graphic illusion of artistic illustration as the result of a deliberate technical-mathematical operation" (Busch, 1995, p. 65). The new combination of eye, hole, painting, mirror, and outer world starts from the eye of the observer and no longer from the eye of God.

But this eye was as unGreek as it was unchristian. For Brunelleschi's image to be developed, it must first have been clear that the inner eye is a darkness into which the light sends its rays, and the pupil at the entrance to this darkness thus functions exactly like the hole in the *camera obscura*. Leonardo da Vinci, whose left-handed manuscripts describe the *camera obscura* in great detail, also articulated this analogy between the *camera obscura* and the pupil (Eder, 1978, p. 39). But through this analogy the eye itself became operationalizable, which means, as always, replaceable. Many observers could hold their eyes up to Brunelleschi's small hole, which also had the form of a conical visual ray. The mirror, the hole, and the painting performed an automatic image analysis for all of them.

The historical break, it seems to me, is that such an automatic image analysis was permissible at all. Under the unshakable theological condition that all creatures were, to varying degrees of exactitude, images of their creator, and that humans in particular were, as the first book of Moses says, *ad imaginem et similitudinem nostram* – created by God in our image (which the biblical plural "our" always implies) – image analysis itself remained forbidden. The ritualistic imperative of image worship prevailed instead, which ruled out the possibility of sending a likeness of God through the hole of the *camera obscura* (never mind the original image itself). The *camera obscura* put an end to this imaginary function, which drove people to recognize or misrecognize themselves only in the likeness of a saint and the saint itself as a likeness of God. In this respect, it was not simply a new scientific device or toy, but rather a weapon in the war of religion. As we know, the media-technical basis of the Reformation was the dismantling of the Bible into printable letters that private individuals were allowed to decipher and interpret without the church making up their minds for them; fathers were even permitted to read the book aloud to their wives, children, and servants. The dismantling of images into portrayable, constructible elements like points, lines, and surfaces similarly brought an end to the painting of icons, and on this so-to-speak clean slate new forms of mathematical analysis emerged, such as Leibniz and Newton's new arithmetic and the geometry of Descartes, the inventor of our coordinate system for planes and spaces.

I would like to point out a third possibility of analysis that the modern age granted to us: namely, the dismantling of flesh and body parts using gunpowder, which became possible only slightly earlier. After all, Roger Bacon, who mentioned the *camera obscura* for the first time, also provided the first correct recipe for gunpowder. And Nicolas Oresme, who replaced the Aristotelian doctrine that all bodies move because they want to return to their natural place with a mathematical analysis of the individual phases of movement of flying bodies – these kinetics should already remind you of film – was a contemporary of Bertold Schwarz, the half-mythical Freiburg monk and inventor of modern guns. Third and finally, as Virilio has repeatedly pointed out, the painters who made essential contributions to the theory and practice of the *camera obscura*, like Dürer or Leonardo da Vinci, at the same time also made essential contributions to the construction of fortresses and the defense of cities against these new guns (Virilio, 1989, pp. 49–50). Dürer's 1527 *Befestigungslehre* (The Theory of Fortification), for example, is a description of perspective from the perspective of ballistics. In other words, the profound aim of the *camera obscura*, which elevated it above many other simply entertaining inventions of that time, converged with the profound aim of shooting, in order to bring down the enemy when he is finally and accurately within one's sights. Together with the new firearms of the modern age, therefore, the *camera obscura* started a revolution of seeing, which was nothing other than the introduction of perspective in general. Humans have painted since the Stone Age, as we know, but it is only since Brunelleschi that these paintings have been based on a constructed central vanishing point to which all the elements of the image refer.

I now come to the second question I posed myself. What made Brunelleschi employ perspective as a mathematically based painting technique rather than as the worldwide spread of divine grace? I have already mentioned how he went from being a craftsman to a mathematician and architect after losing the competition for the design of the bronze doors of the Florence Baptistry. This mathematical, architectonic know-how offers at least a hypothetical reference to the reasons for Brunelleschi's innovation.

In his seminar *The Four Fundamental Concepts of Psychoanalysis*, Jacques Lacan dealt extensively with a topic that Freud mostly neglected: the gaze. I urge you to read the relevant chapter. It is less well known that in his seminar on psychoanalytic *Transference*[3] he also briefly yet dramatically outlines the genesis of linear perspective. Like Hegel, Lacan begins with the hypothesis that the oldest form of art and/or worship was architecture. In contrast to Hegel, however, Lacan makes it clear that there is no god at the center of this architecture, but rather, like the interior of pyramids or temples, there is only a corpse. This corpse needs a space, that is to say, a vacated place, that is to say, exactly like Brunelleschi's image: a hole. Lacan even defines the sacred itself as this architectonic hole: as the presence of an absence.

But every Egyptian pyramid shows – and this is the crucial point in Lacan's argument – how costly the preparation or maintenance of such holes can turn out to be. Millions of stones serve merely to encase a non-place. Lacan conceives of the invention of linear perspective as a simple act of "economy." Instead of building the sacred void, it is much cheaper to paint it as a vanishing point. This artistic innovation has an immediate influence on architecture, according to Lacan, because he conceives of early perspective painting as mural or wall painting rather than panel painting. In Assisi, for example, where the first pre-perspective paintings surfaced and were destroyed in last year's earthquake, murals cover the walls of buildings and thus give them vanishing points or holes that are not actually part of the structure, but are rather cheaper or more imaginary. I will later return to this combination of painting and architecture, like the baroque *trompe l'oeil*.

Unfortunately, Lacan did not know the history of Brunelleschi's hole pattern, which would have confirmed that all perspective painting centers around a hole, and that there is a connection between architecture and painting. Not only is the object seen in Brunelleschi's image a work of architecture – namely, the Florence Baptistry – but it is also the prescribed place from which the illusion of perspective solely becomes apparent – the Cathedral of Florence, as it was finally completed by Brunelleschi's brilliant achievement in dome construction.

---

[3] The German word for transference, *Übertragung*, also means "transmission."

This leads me to the last point of this seemingly never ending commentary on a single image. It concerns the simple question of how Brunelleschi was able to paint his image at all. All of the answers to this question can only remain hypotheses, as Manetti himself did not write a single word about how the painting was done. Even Busch is succinct and resigned: "It is unknown precisely how the production of Brunelleschi's image panel was accomplished" (Busch, 1995, p. 402). However, Shigeru Tsuji, art historian at the Gedei (the Japanese abbreviation for the Imperial Art School of Tokyo), has presented a hypothesis that is so wonderfully plausible I can only endorse it.

Like all good detectives in crime novels, Tsuji begins with the facts in order to question why Brunelleschi chose precisely this image and no other. Why was his image so unusually small (approximately 27 centimeters square)? Why did he paint his image from the main portal of the cathedral? Why was the image obviously painted in reverse, such that only the use of a mirror would make it visually coincide with the reality of the Baptistry? The answer, which resolves all three of these questions at the same time, is that Brunelleschi employed a *camera obscura*. He was therefore the missing link between Roger Bacon in the fourteenth century and Leonardo da Vinci in the sixteenth.

First argument: in Brunelleschi's time there were no lenses. The perforated disk in front of the projected image thus had to be positioned in a place that remained shaded even during the day. This is precisely true of the main portal of Santa Maria del Fiore.

Second argument: the object to be projected must itself lie in direct sunlight. This was precisely true of the Baptistry during the morning.

Third argument: the projection surface must be a certain size. If it was too large the image would become dark and blurred, but if it was too small Brunelleschi's hands would not fit between the perforated disk and the projection surface. With the meticulous use of actual architectonic relationships in Florence and trigonometric functions, Tsuji elegantly shows that Brunelleschi's chosen image size was ideal for his purposes.

Fourth argument: in Brunelleschi's time there were still no geometric devices that could manage to reverse pages automatically. In other words, a reversed image could hardly have been produced by hand in 1425. If Brunelleschi had painted by hand, he could have simply not used the mirror and instead turned the front of his painting towards the observer (rather than the back).

To me, at least, Tsuji's arguments are completely clear. An inventor of a process may thus have been identified only by means of circumstantial evidence, a rare occurrence in the history of media. But

Tsuji rightly emphasizes that even if Brunelleschi actually invented the *camera obscura* as a practical painting device, he did not solve all of the problems of linear perspective painting. The *camera obscura* only works in the real world. This was conclusively proven by its development into the photographic camera, which cannot record anything that does not exist. But the painters of the Quattrocento and the following centuries were very frequently ordered to paint what did not exist: God, saints, and the beauty of earthly rulers. The simple question for Brunelleschi's successors, therefore, was how to take the geometrical automatism of the *camera obscura* and transfer it to other media.

2.1.2.2 Alberti

The only other medium that was possible at that time was paper, which reached Europe from China via Arabia to then revolutionize mathematics, science, and accounting. The problem was how to construct perspectival drawings on paper geometrically, especially when these drawings were pure fantasy or – in the case of new building plans – pure dreams of the future. This problem was first solved by a younger friend and pupil of Brunelleschi's, who attained fame as an engineer-artist and all-purpose inventor: Leon Battista Alberti.

Like Brunelleschi, Alberti certainly also used the magic of the darkroom to astonish the Florentine people. An anonymous biographer recounts beautifully:

> Through painting itself he also produced things that were entirely incredible and unbelievable to spectators, which could be seen through a small opening in a small box. There one caught sight of high mountains and broad landscapes surrounding an immeasurable lake as well as regions so distant that they could not be discerned with the eye. He called these things demonstrations, and they were meant to be seen as natural phenomena rather than paintings. There were two kinds, which he called day demonstrations and night demonstrations. In the latter, one could see Arcturus, the Pleiades, Orion, and other shimmering stars, and the moon rose behind steep cliffs and mountain peaks by the light of the evening stars; in the day demonstrations the shining god was unveiled, who according to Homer was announced far and wide around the world by Eos, the bringer of morning. (quoted in Vasari, 1983, p. 347)

The *camera obscura* can hardly be defined more clearly: it is the sun cult, as for the Greeks – the return of the gods, the enemy of all

Christianity. This was the reason why it was so important to spread the renewed unveiling of being in its entirety, which Helios and/or Alberti achieved, to the world outside Florence. Alberti takes up his pen – Gutenberg had not yet invented his art – and as a grateful pupil dedicates his *Three Books about Painting* to Brunelleschi – first in Italian in 1435 and in scholarly Latin the following year.

The first book of this treatise presents "unheard-of and never-before-seen arts and sciences," which are explicitly without ancient "teachers" (Alberti, 1966, p. 40). To describe linear perspective as a free geometric construction, Alberti developed the concept of an ideal or simply imagined window. This *fenestra aperta* could be considered to be the ancestor of all those graphic user interfaces that have endowed computer screens with so-called windows for the past 20 years. Alberti's window – like Microsoft Windows – was naturally rectangular and could thus be easily broken down into smaller windows. As a model or metaphor for this scanning technique, which was his greatest invention, Alberti employed a semi-transparent veil divided into small rectangles using vertical and horizontal threads of canvas. It could thus be said that in Alberti's work Brunelleschi's single hole became a thousand-eyed Argos. Indeed: Alberti, and later also Dürer, assigned the eye the task of looking through every one of these countless holes into the world of either real models or ideal art objects.

Alberti's real trick, however, was to make even this activity of the eye as virtual as the concept of the window. To do this he used not canvas – the material basis of all painting – but rather paper. The scanned rectangle was transferred out of the world and onto the paper, where it appeared as a checkered pattern, so to speak. This pattern then allowed geometrical constructions to be performed – in other words, operations with Dürer's ruler and compass – to such a high degree of accuracy that the resulting drawing obeyed all the laws of linear perspective. Alberti explicitly emphasized that he had written his treatise for artists and not for mathematicians, which is already clear from the title. For this reason, as I have implied, the applied mathematics still adhered to the good old Euclidean proportions between lines and angles. In other words, it did not look for help from the new trigonometric tables. Even more gratifying and enigmatic is the historical fact that Regiomontanus, the creator of the best trigonometric tables, undertook a trip to Italy, and during this trip – in Ferrara – he reportedly met Alberti. I would be a happier man if I knew what they had talked about.

Not knowing this is one of the reasons why a simple historical question cannot be completely clarified: what was the practical cause of this radical shift in the fifteenth century – from the two-dimensional miniature to the perspective panel, from the pictorial nature of all God's creatures to the mechanics of the *camera obscura*?

It hardly needs explaining why it was necessary to learn to see in perspective when shooting, whose invention I previously alluded to. The reason why it was necessary for painters to learn to see in perspective following Brunelleschi's experiment, however, was previously attributed by art historians to a *Stilwillen* – or "will to style" – that simply led to the new Renaissance art. A better explanation is already implied by the fact that in the very beginning, experiments with the *camera obscura* could only be conducted in darkened yet otherwise normal-sized chambers or rooms, but they soon changed to become small, transportable boxes. (Consider the difference between literally fixed temples and transportable Bibles.) Painters who had a *camera obscura* could thus "paint according to nature," as the lovely phrase goes, simply because the small, portable box allowed the light and everything it illuminated to be conveyed onto a surface, which the hand of the painter then only had to paint over. People have always painted according to nature in some way, just as when the puppeteers in Plato's allegory of the cave produced silhouettes of jugs and similar tools, but they have not always made the hands of the painter into dependent functions in an experimental procedure. As if anticipating Arnheim's theory of photography, on the other hand, the *camera obscura* combines for the first time the optical transmission of information with the optical storage of information; the former function is already fully automatic, whereas the latter is still manual.

We will not dwell on this manual limitation, but rather we will stress that the number of drawings and images generated with the aid of a *camera obscura* is probably beyond the wildest dreams of a hermeneutic history of art. The benefits are obvious: the drawings that result from this union of optical receiver and human data sink, *camera obscura* and painter, naturally have a greater level of precision. This precision also became, as in Dürer's work, a theme of triumphant and self-referential drawings, which then once again recorded (for educational purposes) how the painter captures the image of a woman on paper either through a lattice placed in the room or by way of a *camera obscura*. I will only point out here that it goes without saying that women were once more the subjects of such experiments, but since this is a media history and not a love story, I prefer to steer clear of my suspicions concerning the purpose of the whole episode.

As I said, we do not know whether Alberti spoke with Regio-montanus about trigonometry and linear perspective at the court of Ferrara, but we do know the content of another conversation that Alberti had in his old age. This conversation has come down to us from Alberti himself, and it gives unexpected information about the causes that drove the modernization of technical media in the middle of the fifteenth century. In 1462 or 1463 – we do not know exactly – Leon Battista Alberti took a stroll in, of all places, the Vatican gardens with, of all people, a certain Dato, who was by profession secret scribe to the Pope. I should explain that the field of encoding and decoding texts, which began in the ancient world, was to some degree neglected in the Middle Ages. Cryptographic specialists were only employed in the Vatican and by the *Signoria* in Venice, where modern diplomacy in general originated. Dato, with his absolutely appropriate name, whose plural is "data," was one of them.

Alberti opened the conversation quite differently. He said that while an hour of chatting was spent in the Vatican garden, the "man in Mainz" had probably made another dozen or hundred copies of a rare and irreplaceable manuscript of ancient knowledge by laying it under his printing press. In other words, Alberti explicitly saw himself as a contemporary of Gutenberg. Dato must have answered – no one knows for sure – that in spite of all the Gutenbergs of this world, cryptanalytic encoding, his own profession, unfortunately is and remains a lengthy process.

It seems to me that this complaint preyed on Alberti's mind. He immediately sat down, with a quill in hand naturally, and thought about how the process of encoding and decoding secret messages could be accelerated, just as Gutenberg's movable type had accelerated handwriting or made it entirely superfluous. What emerged was a treatise on ciphers, which continues to be the basis of all cryptography, even in the computer age, as David Kahn, the leading historian of cryptography, emphasizes.

Albert introduced two innovations. One, strictly according to Shannon, on the side of the sender, the other, again strictly according to Shannon, on the side of the receiver. When Roman emperors like Caesar or Augustus encrypted their messages, they simply moved all the letters one or two places further along in the alphabet, although Augustus never mastered modulus mathematics and therefore did not code the last letter X as the letter A (Suetonious, 1979, pp. 39 and 102). It was quick but also easy to crack. Alberti transferred the principle of movable type from Gutenberg's printing press to cryptography. Whenever a letter was shifted alphabetically and then

64

written down according to the code, the code itself also changed. The next letter to appear on the paper was shifted one additional place in comparison with the original text. This remains the basic principle of polyalphabetic ciphers today.

Alberti's innovation in the field of decryption was decidedly Gutenbergian. The printing press had already made it plain that in order to print normal texts many more E's were needed than, say, X's or Y's. A glance in any typesetter's case will confirm this. Alberti, like Edgar Allan Poe's *X-ing a Paragrab*, threw precisely this glance at texts encrypted in the old-fashioned, manual way and not through his polyalphabetic method. When there are far more Y's than E's in such a text, this means plainly and simply that the letter E has presumably been encrypted as Y. In other words, Alberti transferred the coldness of numbers to the sacred realm of everyday grammatical sense or semantics.

2.1.3 Impact

2.1.3.1 Perspective and Letterpress

This long digression into the history of textual media should make one thing clear: Alberti mathematized old manual techniques like painting and writing, and at the very least he had explicitly made reference to this modernization before Gutenberg. The question remains whether this reference before and to Gutenberg is not also true of Alberti's mathematization of painting. Busch cites a remarkable passage, though I have not been able to verify it. No less a person than Giorgio Vasari, the contemporary and biographer of all of these painters, wrote in his 1550 book *Lives of the Most Eminent Painters, Sculptors and Architects* that "in the year 1457, when the very useful method of printing books was invented by Johann Guttenberg, a German, Leon Battista discovered something similar," albeit merely in the field of painting (Vasari, 1983, pp. 346–7). In an age of growing national pride, this was probably supposed to imply that Italy's technical achievements had caught up with Germany's. Contemporaries thus already saw a connection between the art of artistic writing and the art of artistic perspective, *ars artificialiter scribendi* and *perspectiva artificiosa*. This supposition can be theoretically substantiated.

The content of a medium, McLuhan decreed, is always another medium. All of the Renaissance drawings, which described how to build a *camera obscura* and how best to install it between the painter and the living object, were stored and passed down in books, particularly in textbooks. For the first, yet certainly not the last time, we are encountering something like a union of media: the printed book, on the one hand, and the drawing brought to a higher level of precision through the *camera obscura* or linear-perspectival geometry on the other. As soon as one recognizes that, the lowest common multiple of the two media becomes obvious. Through Gutenberg's invention it was possible for the first time that all of the copies of a book, or at least of an edition, presented the same text, the same printing errors, and the same page numbers. As Hans Magnus Enzensberger wrote in a poem about Gutenberg, "How greatly this page here resembles a thousand other pages" (Enzensberger, 1976, p. 4). (Not to mention the uniformity of computer software, with which my lecture notes and the notes of countless others have been drawn up.)

Elizabeth Eisenstein very convincingly argues that the new, mechanically perfect reproducibility of the medium of handwriting also put competitive pressure on other manual arts. The reproducible book as such required illustrations that were equally as reproducible and exact – not to make readers or art lovers happy, but rather to store and transmit technical knowledge, the most shining example of which was the invention of the letterpress itself. Eisenstein directly connects the great upturn in technology, science, and engineering in Europe in the modern era with the availability of technical drawings, construction plans, and sketches, which looked the same in every printed copy simply because they were indestructible reproductions of a single original. As we know, the techniques of wood engraving and copperplate etching, which were developed or perfected at that time, provided this reproducibility, whose lack in other cultures resulted in drawings showing more mistakes – or more noise – as they were copied from copies of copies, etc. But who or what ensured that the original was a correct reproduction of its original, which may have been a woman or the *camera obscura* itself? My supposition: scientifically based perspective and its technical implementation – in other words, none other than the *camera obscura* once again. Even though the *camera obscura* was not a camera in the sense of photography or film, and consequently it could not replace the manual work of drawing and painting, these handicrafts nevertheless fell under its scientific-technical control. When one realizes that in the centuries before Gutenberg's invention the operational secrets of all manual workers were always only communicated from master to journeyman, from generation to generation, and when one realizes that secrecy was so important and promising, that entire cults and rituals

were erected around it (like wrought-iron work), one can appreciate what it means to be able to entrust building plans along with
explanatory texts henceforth to the printed book. Real guild secrets
were replaced by the knowledge of engineers, which was in principle
also possibly autodidactic, and ritual guild secrets were replaced by
the specially invented and complementary secrets of associations like
the Freemasons, which made imaginary theories out of the former
practices of masons.

Print technology made the autodidact possible – that is the point
upon which everything depends. The book became a medium in which
technical innovations as such could take place. They could be stored,
shared, and even advanced with the help of technical drawings in the
text. Models of a mill or a *camera obscura* are easier to understand
than their so-called reality. This is the reason why the excursion into
letterpress was not a digression, but rather it furnished the historical
foundations for the astonishing and otherwise inexplicable fact that
Europe, in contrast to other cultures, has produced one technical
medium after another since the Renaissance. It can concisely be said
that Gutenberg's letterpress made the techniques that superseded it –
from photography to the computer – possible in the first place. It was
the unique medium that set other media free. This is true for Dürer's
age as well as today. Without specifications, manuals, and technical
drawings new generations of computers would be an impossibility.

2.1.3.2 The Self-Printing of Nature

There is evidence to support this hypothesis about the practical uses
of linear perspective and the *camera obscura*. The first piece of evidence also brings up an important detail from the prehistory of photography. Namely, the seventeenth century had already attempted
to eliminate the great handicap of the *camera obscura*, that is, the
necessity of manually painting over the images that emerged. Anato-
mists like Vesalius in Bologna or botanists like Gessner in Basel took
on the epoch-making task of pouring everything knowable about
the human body or the plant world into printer's ink and printing,
which greatly increased the need for precise illustrations. From 1657
onwards, therefore, nature researchers have also experimented with
the possibility of transferring the objects of their research onto paper
without the mediation of a wood or copperplate engraver. A Dane
named Walgenstein, who will soon be mentioned again in connection with the *lanterna magica*, reportedly succeeded in preparing the
leaves of plants so that an imprint of them could be made. At the

start, the leaves were simply held in smoke until they were black enough to leave behind an impression, but later on the very same material employed in the printing of Gutenberg's letters was also employed to print objects, as the leaves were prepared with printer's ink. In any case, the images emerged in their natural size and with all the detail, but unfortunately only relatively few Gutenberg leaves could be produced from one botanical leaf. After that, it was worn out and had to be replaced by another leaf. Such attempts at least show, as Eder has already emphasized in his lengthy and old *History of Photography*, the clear tendency to set technically reproducible scientific illustrations alongside technically reproducible type (Eder, 1978, p. 33) – not only, as Eder assumes, to save the high costs of copperplate and wood engravers, but also to be able to compete with the precision, and that means from that time on the scientific nature of reproductions. This clarifies the connection between perspective representation, the *camera obscura*, and Gutenberg technology. In short, we can say that leaves (of plants) became leaves (of books) – while plants of the field, forest, and meadow became the content of optical media.

2.1.3.3   Europe's Colonial Power

The second piece of historical evidence is even more amusing or eloquent, at least for people who do not suffer from political correctness. In his wonderful book, *The Heritage of Giotto's Geometry*, Samuel Edgerton also recounts the history of the Jesuits who invaded China in droves during the Ming Dynasty, starting around 1600, in order to preach their faith – and not without success. The reason why the missionaries belonged to the Jesuit Order of all people is still being considered today.

In Peking, Father Matteo Ricci and his successors started an enormous enlightenment campaign. They equipped their own library with scientific books, and would you believe it, 19 of these titles were about perspective (Edgerton, 1991, p. 261). Rather than educating the heathens, though, the Jesuits planned to convert them by producing and distributing Christian images, which had already helped an otherwise very insensible theology to triumph in Central and South America. However, the Jesuit in the Vatican who was responsible made a bad mistake. He determined that the native draughtsmen and copperplate engravers who were supposed to translate the images of Christianity into the image universe of Chinese culture should not be trained in Peking itself, but rather in distant Japan (Edgerton,

1991, p. 266). In other words, the 19 treatises on perspective, from which the so-called natives could have learned to draw, were not available.

Thus it happened as it had to happen. In 1627, Father Johannes Adam Schall von Bell decided to send four ambitious volumes with "diagrams and explanations of curious machines from the Far West" to the printing presses in Peking (Edgerton, 1991, p. 271). The so-called *Theatra Machinarum*, a book genre that not coincidentally had flourished since the Renaissance, normally contained exact perspective copperplate or wood engravings of existing or merely fictional machines – sketches, therefore, which supposedly enabled the observer to successfully recreate three-dimensional machines from two-dimensional images. Schall's native, presumably Japanese, wood engravers accordingly went to work. They had the European books along with a Chinese translation of the texts directly in front of them, but they were nonetheless completely incapable of correctly copying the proportions in perspective.

Up until the first decades of the nineteenth century, imperial China continued to print these kinds of incorrect graphics in encyclopaedias and scientific-technical manuals. You can imagine the results. China was the most technologically advanced country in the world during the Middle Ages, but it remained trapped in a state that made it very easy for the English and other European powers to defeat China in one war after another from 1840 onwards. Perhaps the lesson to be learned from this is that linear perspective was not simply an aesthetic or artistic shift in taste, but rather a thoroughly technical re-evaluation of all optical values, which was inconceivable without the corresponding mathematical qualifications, such as during the Ming and Manchu dynasties in China. In a story by E.T.A. Hoffmann, which I will return to later, a "Chinaman" of all people poses the "stupid question": "How is it that objects grow smaller as they recede?" (Hoffmann, 1952, p. 77). Linear perspective remained one of the arcana of modern European power until approximately 1850, when it once again reached Japan and elsewhere.

So much for linear perspective from the perspective of what Shannon calls the receiver side. The *camera obscura* captured light and cast it further, but it did not send it. For thousands of years, that was left entirely up to simple signal systems consisting of mirrors and torches, which would determine the outcome of battles. Long before Einstein's proof that the speed of light could not be surpassed, soldiers already knew the advantage of rapid communication.

2.2 *Lanterna Magica* and the Age of the World Picture

The last task of this brief history of art must therefore be to recount when, how, and why images also received a transmitting apparatus and thus learned how to be transmitted without the classical means of transportation provided by the postal system. The transmitting apparatus was a technical but not a historical twin of the *camera obscura*, and it went by the lovely name *lanterna magica* or "magic lantern."

In essence, the *lanterna magica* simply turns the *camera obscura* inside out. A hole in a wall once again separates inside and outside, system and environment. But in place of the sun, which in the *camera obscura* transmits images from the environment into the system, the *lanterna magica* employs an artificial light source in the interior of the system, such as a simple candle. Using either front or rear projection, this candle shines through interposed concave mirrors, or later systems of lenses, and illuminates a drawn and often colored pattern, whose mirror image is then projected outside through the hole and onto a screen – the forerunner of all film screens. So much for the principle, now for the history.

2.2.1 *Magic Lanterns in Action*

The direct precursor of the magical device was the well-known bull's eye lantern, which was made by Liesegang, a photography dealer and historian in whose honor his hometown of Düsseldorf happily renamed a street. This ancestral line appropriately casts the light of war on the *lanterna magica* (following Michel Foucault). Bull's eye lanterns were officially used to illuminate battlegrounds, but they were unofficially used by hunters, fishermen, poachers, and murderers. Even today, Greek fishing boats sail out in the Aegean at night using bull's eye lanterns to lure their prey like moths into an inescapable light-trap. Bull's eye lanterns were absolutely forbidden during Absolutism for similar reasons: in German principalities they were forbidden because poachers (long before the invention of highways, automobile headlights, and dead game across the asphalt) ruined the route to hares and deer, which belonged to the sovereign, who had an absolute hunting monopoly; in France, the punishment for using bull's eye lanterns was the death penalty, because murderers could use them to paralyze their victims like a snake. The problem of fixing movement was therefore virulent long before photography, and the purpose of deploying bull's eye lanterns on the battleground was not only to make friendly movement easier but also to make enemy movement impossible.

The *lanterna magica*, which was presumably the descendant of such lanterns, was unfortunately faced with an entirely different problem. It was not supposed to make movement impossible, but rather to simulate it. While the *camera obscura* had helped to project images, even images of moving objects, the *lanterna magica* did the exact opposite. An image of the object was moved in front of the lens system and a mirrored light source produced an enlargement, which naturally seemed considerably more alive or threatening. There were reportedly projection mechanisms that could display (believe it or not) 12 images at once; where a single soldier had done his drill at the information source, 12 soldiers (as you can easily work out) performed their well-trained threatening gestures in step at the information sinks. The strategic techniques developed for bull's eye lanterns, which deployed naked light devoid of any specific form or shape, became imaginary techniques of control through the use of images or figures. With the exception of the more modern mirror or lens system, therefore, the *lanterna magica* is a reversal of the *camera obscura*. That may be the reason why earlier historians attributed its invention to the same Renaissance researchers to whom the *camera obscura* can also be traced back. But in the meantime, it has been proven that Giambattista della Porta did not have access to a functioning *lanterna magica*. Magic lanterns did not surface until a century later, in 1659, when their (in retrospect) unbelievable career began. There is a scientific-technical reason for this delay – like all optical media even today, they required the development of usable lens systems – and there is also a second reason, which is worth discussing. The second reason leads to magic and conjuring, and it delivers us, like Goethe's Faust, from the dust of the lectern.

2.2.2 Implementation

It appears to be no accident that the development of the *lanterna magica* was not attributed to artists and painters, like the *camera obscura*, but rather to two mathematicians: besides the Dane Thomas Walgenstein, who reportedly demonstrated the self-printing of nature from the leaves of plants, as mentioned above, it was also attributed to the great Dutch mathematician, physicist, and astronomer Christian Huygens. Walgenstein presumably studied with Christian Huygens at the University of Leyden, which was famous at that time, and he reportedly said that he took a "bagatelle," which Huygens had

71

not pursued any further, and made it effective and ready to go into production (Schmitz, 1981–95, p. 294). Indeed, Huygens not only described the wave theory of light, without the knowledge of which televisions would not function, but he also made practical improvements to optical lens systems and built one of the first usable celestial telescopes. You can already guess the results of this arming of the eye with glasses and lenses, telescopes and microscopes in the seventeenth century: the postulate of the visibility of all things collapsed under the evidence of the invisibly small under the microscope – like spermatozoa – and the invisibly large through the telescope – like the phases of Venus or the rings of Saturn. As we know, half of Pascal's philosophy was concerned with this, as well as the entire mathematics of differentials and integrals, which Leibniz invented while studying with Huygens of all people.

The impact of lens systems on everyday life also becomes evident when one realizes that in imperial Rome only one person had access – namely, the emperor himself, according to the near-sighted Nero – not to glasses, but rather to a piece of emerald, which was formed in such a way that it took the place of glasses at gladiator games. In short, it could be said that the baroque technology of lenses forced physical light itself, with its optical paths and refraction indexes, into the perspective that was invented only theoretically in the Renaissance. Huygens did not deal with both reflection and refraction without reason in his *Traité de la lumière*, for optical media like the *camera obscura* and the *lanterna magica* implied a considerable increase in image definition: the primitive hole, which only prevented blurring in a negative way, namely by filtering, but could never become the ideal, namely an infinitely small hole, was replaced by the positive possibility of gathering and concentrating light. It was no wonder, therefore, that both these optical devices were applied on a massive scale following the development of lens systems, and they soon surfaced in such different areas as science, art, and religion, as well as in magic and folk entertainment.

2.2.3 Impact

2.2.3.1 Propaganda

As in the case of the *camera obscura*, it is also only reasonable not to attribute the mass application of the *lanterna magica* in the following centuries simply to linear scientific-technical progress. It is important to note that the first reports of its deployment were not for scientific purposes, but rather for the purpose of creating illusions. For example, the scientist Huygens reportedly refused to build a bagatelle like the *lanterna magica* for his father, who was a famous writer (Schmitz, 1981–95, p. 294), and his mathematics student Walgenstein reportedly employed the *lanterna magica* not to conduct research, but rather to spread fear and horror among his spectators by projecting death as a skeleton.

It was this ghostly use of the *camera obscura* that made careers in the following period. Its deployment was logically first considered by writers who were scientifically interested, but who were above all religious soldiers and Catholics, such as the Jesuits Athanasius Kircher and Kaspar Schott and the Premonstrant monk Johann Zahn, who reportedly built hundreds of witch lanterns (Zglinicki, 1979, p. 51). The magic in the name of the device, which could feign things to the eyes that not only happened to be absent but could also never be present, like ghosts, thus likewise needs derivation.

To bring the matter into focus, I will limit myself to Kircher, who was also one of the most exciting figures of his time. Athanasius Kircher came from around the region of Fulda in Germany, joined the Jesuit Order when he was very young (as was customary), became a professor of philosophy and mathematics (a combination that was still possible or even typical at the time), left Germany during the turmoil of the Thirty Years' War, and ended up in Rome at the Vatican. Apparently, the Holy See decided at that time to avoid any scholarly scandals in the future, like Galileo or Giordano Bruno; in any case, Kircher rose to become a kind of scientific fire brigade for the Pope: with a special mandate and special clearance he was always present when there was new scientific territory to explore as well as defend in the name of the church. Kircher's publications ranged, logically enough, from mathematical combinatorics to Greek mythology to the alleged decoding of Egyptian hieroglyphics. He was a classical case of polyhistory, as it was called in the seventeenth century.

A hint concerning the magic of the *lanterna magica* and the purpose of its deployment can already be found in the title of the magnificent volume, in whose second edition in 1671 Athanasius Kircher presented a sketch of the *lanterna magica*, although it was not entirely technically correct. The book is called *Ars magna lucis et umbrae*, the great art of light and shadow, which implies that the new optics was employed not as an instrument for scientific research but rather as an art. Among Kircher's arts, which clearly have nothing to do with Kant's aesthetics, two stand out: one military and one religious.

73

It was Kircher's hope that a modified form of the *lanterna magica* could revolutionize signalling or communications during wartime. The use of simple torches since the time of the Greeks had always given generals the greatest difficulties in transmitting orders that went beyond mere oppositions like yes and no, light and dark. In other words, they could not send coded alphabetical messages. However, Kircher proposed a concave mirror with written symbols corresponding to the order to be transmitted, whose letters could be blinded. The mirror could then be held in sunlight and an articulated message could be transmitted in this way over a distance of up to 12,000 feet or three and a half kilometers, without potential enemies within this distance having a chance of intercepting or actually hindering the transmission. Kircher's signal system project was thus called *cryptologia* or *stenographia* – secret writing with light – more than a century before Claude Chappe's optical telegraph.

As Liesegang's history of photography already noted, entertainment media like the *lanterna magica* were not developed for entertainment purposes, but rather they were byproducts or waste products of pure military research. In the age of intercontinental ballistic missiles and Teflon pans, one would say spin-offs. It is also not surprising that Kircher belonged to the only order of monks that had and still has a general at its head. As Liesegang puts it, he was looking for a telegraph communication system that was supposed to guarantee perfectly secret communication between the members of a militant elite, and instead he popularized an optical medium of entertainment, and the more uninformed and greater in numbers its observers were, the more impressive or magical it seemed. The same transition from the telegraph to simulation, from the symbolic to the imaginary, will return with Edison's invention of the phonograph and the kinetoscope.

The logic of replacement or spin-off raises the question as to whether and how a strategic plan remains valid in the technology of the imaginary. To answer this question, Kircher's second, religious artwork must be presented. Namely, the Jesuit priest proposed a device that was the direct precursor of the zoetrope and must therefore be regarded as the direct precursor of film: the so-called "parastatic smicroscope" (Zglinicki, 1979, p. 56). As its name suggests, this device displayed or juxtaposed (those are the definitions of "parastatic") very small things, exactly like the microscope, and it consisted of a turntable and an optical observation facility. Small images were placed on the turntable, and enlargements could be seen through a lens system. Only Kircher's images were not simply dancers and models that were able to turn around and move, like the images later used with the stroboscope in 1830, but rather they were the famous stations of the Passion of Jesus Christ, which had been established for centuries. These stations were earlier painted in spatial and temporal succession in churches or along the Way of the Cross, such as St. Ottilien, but Kircher began to reel them off at time-lapse speed and they began, if you will, to move. The Mount of Olives, Calvary, etc. as the first silent film in the history of media . . .

The obvious question here must be why the seventeenth century did not denounce Kircher's suggestion as blasphemy, why it was permitted for the first time to show the central concept of the Christian message as a visual work of art, that is, as an illusion of a witch lantern.

The answer, I suspect, lies in the concept of optical transmission itself. Both of Kircher's proposals have a common goal: to send optical information and thus produce military or religious effects among the receivers.

2.2.3.2 Heidegger's Age of the World Picture

There is also the question of what made the *lanterna magica* so attractive for Jesuits of all people, but before I answer this question I want to make a short diversion into the realms of philosophy. As we know, in his late work Martin Heidegger attempted to think of the basic concept of European philosophy – being – as historically changeable despite all tradition. According to one of Heidegger's theses, being first constituted itself in the form of a representation (*Vorstellung*) in European modernity. Representational thinking delivered being as an object for a subject, which was not at all true for the Greeks and the Romans (Heidegger, 1977, pp. 132–3). A lecture on optical media can verify the facts of the case: it can be said, following Heidegger's line of thought, that linear perspective and the *camera obscura* were precisely the media of this representation.

In the next step in the history of being – in the philosophy of Descartes – the representation of the subject is re-presented to the subject once again as such: *cogito ergo sum* – I am because I can represent anything presented before me. As Descartes makes very clear, in the *cogito* the difference between day and night, waking and dreaming, reality and hallucination does not count. This also allows us to grasp the history of optical media more precisely: the technical device that re-presents representation itself (instead of reality) is of course the *lanterna magica*. The image of something – in other words,

75

its representation – is slid into the black box, and it is illuminated by a light that casts a representation of this representation, an image of this image, onto the wall. That concludes the history of being and my explanation of why the *lanterna magica* could not have come into existence until 100 years after the *camera obscura*.

2.2.3.3 Jesuits and Optical Media

But let us return from thinking back to theology. The *camera obscura* was directly linked to the letterpress. At least indirectly then, the *camera obscura*, linear perspective, and the Reformation went together – if for no other reason than because Luther's precept by which Protestant Christianity was founded on pure faith and pure writing could, on his own admission, not have been technically implemented without Gutenberg. You may have guessed, however, that this modern precept would not exactly be embraced by the one true church. This is why the situation called for counter-measures to arm the old faith technically.

A short time before this, in 1622, Pope Gregor XV set up the *congregatio de propaganda fidei* in the Vatican: the association for the dissemination or propaganda of Catholic belief, which was the first propaganda agency in history. In the same year, the same pope canonized the founder of the Jesuit Order. A few years later, the attempt to propagate linear perspective in Peking failed. No history of optical media should hide the fact that entertainment media are always also propaganda machines. But above all, there exists in every epoch of optical media good reason to name their strategic relationship to the enemy, that is, the written word. The only thing that occurs to Zglinicki concerning Kircher's smicroscope is the involuntarily comical sentence "we clearly see here once again the aim of the world at that time to visualize the events with which one occupied oneself" (Zglinicki, 1979, p. 56), yet such tautologies must first be purged from a history of media war. Does the innocent "one" possibly have a darker proper name? What was the optical implementation of 6, 8, 14, or 36 Stations of the Cross about?

In at least half of Europe, the Reformation had abolished or literally blackened medieval church rituals, with all of their visual glitter, and replaced them with the monochromatic, namely black-and-white mystery of printed letters. *Sola scriptura, sola fidei* – solely from writing and solely from belief. According to Luther, therefore, everyone should be able to be blessed without the worship of holy images and doing useful works for the church. Or, as King Crimson sings: "Starless and bible black." The rest of Europe had to take some action against this bible black, and it invented, as you know, the so-called Counter-Reformation, which meant above all religious propaganda and the Jesuit Order, to which *lanterna magica* practitioners Athanasius Kircher and Kaspar Schott belonged as well as the perspective propagandist Father Schall.

The Jesuit Order was the work of a single man, who was not coincidentally a soldier by profession. Inigo Lopez de Recalde, later known as Loyola, was the scion of a Basque squiredom, and in 1521 he defended Pamplona against the French. He thus fought in a fortress designed by those artist-engineers, which was systematically destroyed by the cannons designed by the self-same artist-engineers. It so happened that a cannonball badly wounded Loyola's right leg, and his prolonged recovery from this war wound turned into a religious conversion. Loyola began (probably for the first time) to read books, consumed one holy legend after the other, became increasingly religious and eventually dedicated the tools of his trade – his weapons and armour – to the miracle-working image of the Virgin Mary in the monastery of Montserrat.

Fever and delirium, books and holy images – that was more or less the medial context from which the Jesuit Order emerged. The *Exercitia spiritualia* of holy Ignatius, the founding text of the order's founder, was already a book opposed to all books. For Loyola, spiritual exercises meant exercises for the soul as well as the body, which army reformers like Maurice of Nassau, Prince of Orange, christened as drill practice around 1600 and which early twentieth-century film theorists eventually called psychotechnics. But unlike Luther and the countless Protestant housefathers who went to Luther's school, these exercises did not consist of transcribing and reading the Bible or throwing the inkpot at the devil, should he want to disturb their writing work. For Loyola, who had been a soldier, drill practice had nothing at all to do with writing, and he only became interested in reading when he was a critically ill patient. At some point, the eyes of the founder of the Order or the eyes of those students who, according to him, were supposed to endure his spiritual exercises surely came across the letters of a religious book and noted, for example, what was written about hell. (In all Loyola commentaries, from James Joyce to Roland Barthes to myself, hell is naturally the dramatically preferred example, but in light of Athanasius Kircher's smicroscope the passion play could just as easily be used.) When Loyola or one of his Jesuit pupils locked himself in a monastic cell for weeks in order to meditate on hell, therefore, the intertwined legends were

77

actually involved, but not the book called the Bible. In other words, the scant information about hell, as it stands in the only true book of the Protestants, was always already overgrown and embellished by religious fantasies. For the Jesuit Order, it was logically important to visualize long and intensively everything that had once been read until it stopped being letter or text and began instead to overwhelm the five senses themselves. Let us read what Loyola and his pupils did in the fifth exercise. The set topic read: *A Meditation on Hell.* Here is the run-through:

> *First heading.* To see in imagination those enormous fires, and the souls, as it were, with bodies of fire.
> *Second heading.* To hear in imagination the shrieks and groans and the blasphemous shouts against Christ our Lord and all the saints.
> *Third heading.* To smell in imagination the fumes of sulfur and the stench of filth and corruption.
> *Fourth heading.* To taste in imagination all the bitterness of tears and melancholy and a gnawing conscience.
> *Fifth heading.* To feel in imagination the heat of the flames that play on and burn the [damned] souls. (Loyola, 1963, p. 36)

This means that only someone who had personally completed all these hellish spiritual exercises, like Loyola, could and would be permitted to join the Jesuit Order, and above all they must have imagined the torments of the damned Protestants in their hell-hole right up to their bitter end. It was thus almost too obvious what the Counter-Reformation had to offer in reply to the new Protestant medium of the letterpress: a theater of illusions for all five senses (although the sense of vision took absolute priority in all of the spiritual exercises) and a reading practice for readers who did not stick to the letter but rather experienced its meaning immediately as a sensual hallucination. In other words, the search for a medium that could combat Luther's Bible brought back the old religious images in a changed or improved form – no longer as icons or panels on a church wall, no longer as religious miniatures of the Acts of the Saints that even a child could comprehend, but rather as psychedelic visions that could motivate the soldiers of Christ, as the Jesuits called themselves, in the religious war much more effectively, and that means unconsciously, than the old-fashioned painted masterpieces. On the other hand, as you well know, with regard to even the most devout techniques resulting in ecstasy the established church in Rome has been and still is easily and justifiably suspicious of heresy. It is

78

therefore no wonder that in his early days in Spain Loyola landed in the prisons of the Inquisition several times. It is even less surprising that at the last minute, which means after 1540, when the Pope ordered him to come from Spain to Rome in order to have his drills examined, Loyola made or rather had to make his spiritual exercises compatible with the old-fashioned worship of panels. The sixth and eighth rules of these Roman supplements read as follows:

> We should approve of relics of the saints, showing reverence for them and praying to the saints themselves; visits to Station churches, pilgrimages, indulgences, jubilees, Crusade bulls, the lighting of candles in churches should all be commended [. . .] We should praise church decoration and architecture, as well as statues, which we should venerate in view of what they portray. (Loyola, 1963, pp. 120–1)

We thus come to the end of a short excursion into church history, and find ourselves back with the Jesuit Athanasius Kircher and his optical phase model of the Stations of the Cross – though not without shedding new light on the *lanterna magica*. Loyola's imperative "praise!" established a new kind of image worship, which, like the new hallucinatory readings, was not directed at the image, but rather at its meaning. It was a kind of image worship, therefore, that could not stand by and simply leave the outward appearance of churches or the Way of the Cross as the architects and painters had arranged them, but rather aimed at transferring the psychedelic effect of the spiritual exercises to the outside. "Outside" here refers to many possibilities: first, the outside of a real projection surface on which the inner image could appear, such as the harmless example of panel painting, or the exciting and innovative case of the *lanterna magica* screen. Second, "outside" also refers to the outside beyond the Jesuit Order, a monastic elite whose members had worked over weeks and months with all possible mortifications of the flesh to actually achieve hallucinations. The Jesuits' task, to beat the Reformation's letterpress monopoly with more effective media technologies, was necessarily aimed at the conversion of the lay public, who through time constraints were not expected to have performed spiritual exercises in the cloister. What Loyola had invented or enforced in his lonely cell had to become simplified, trivialized, mechanized, and mass applied. That is the entire difference between a spiritual exercise, in which Loyola hallucinated the Stations of the Cross, and an optical device called the smicroscope, with which Kircher showed rapidly changing images of these stations to the lay public, whom they wished to convert through the precursors of film. The Counter-Reformation triumphed in Austria, Bohemia, Moravia, Silesia, and half of southern Germany not only because of the millions of deaths caused by the Thirty Years' War, and therefore not only because of the dark or negative sides of power; it was just as much due to their bright and that means visual aspects, to a new kind of imaging.

This image technology is not only represented in the title of the second edition of Kircher's *Ars magna lucis et umbrae*, which does not promise to make scientific progress with light and shadow but rather to perform great arts. It can also be found in the illustration of a *lanterna magica*: one sees an oil lamp, and in front of the oil lamp is a horizontal row of painted glass plates that are all waiting to be shown and projected one after the other (almost like film) in the beam of light; but above all, one sees on the dark facing wall the projection effect of the plate that currently lies in the oil lamp's beam of light: a naked man surrounded by waist-high flames. It would not be wrong to assume that these flames only signal to our modern eyes, which have been trained by McLuhan, that the message of every medium is the medium itself – in this case, the oil lamp – but among Kircher's contemporaries and audiences these flames meant something entirely different – namely, the flames of hell. In other words, thanks to the *lanterna magica* the solitary hallucination of the founder of the Jesuit Order, who once concentrated all his five senses on imagining the agonies of hell, became technologically simulated for the masses. And the fate awaiting those who failed to find their way back to the only true faith immediately after the presentation was not simply the projected flames of hell. Kircher writes: "The images and shadows presented in dark rooms are much more frightening than those made by the sun. Through this art, godless people could easily be prevented from committing many vices / if the devil's image is cast onto the mirror and projected into a dark place." (quoted in Ranke, 1982, p. 17). The fact that my source, Winfried Ranke, rejects the conclusion that "this was the beginning of a didactic and perhaps even missionary-indoctrinating deployment of the new demonstration device" (ibid.), despite all of Schott's assurances about the many Jesuits engaged in *lanterna magica* experiments, is thus only evidence of blindness.

In our current mania over film and television, this strategy of effects may seem strange for a new optical medium, but it is necessary to realize that the first great entertainment media of the modern era, theater and literature, arose from a centuries-long war over religion and images. That goes for first-rate cultural performances, which are still retained in historical memory, as well as the lost performances of traveling players, who around 1900 were literally devoured and relegated to oblivion by the technical medium of film.

2.2.3.4 Traveling Players

To start with the simpler category, because I don't need to know anything about their origin and sociology, the so-called traveling players always took up the escalations of the great religious war astonishingly quickly and systematically. When Gutenberg and later Luther appeared to triumph over the worship of images with their printed books, the traveling players discovered the handbill as a medium for reproducing their street ballads and illustrating them with wood engravings. After the Counter-Reformation of hallucinated and projected images came to power, on the other hand, the traveling players seem to have gone over to Kircher with all flags flying. In any case, his smicroscope became the immediate precursor of all those peep show cabinets that were admired at fairs on a massive scale up until the nineteenth century. Peep show cabinets were small, transportable devices for presenting images, which had one or two holes for the eyes and which were often fastened to the backs of traveling entertainers. For a modest fee, people were permitted to look into the box, just as people later looked into the kinetoscope, the immediate precursor of our cinema, and they were rewarded with images that could be mechanically wound on one after the other. Refined models even contained a miniature stage on which the scenery and the dramatic figures could each be moved separately so that a rudimentary narrative was allowed to develop. And when the peep show cabinets ran at the same speed as a mechanical music box, all that was missing was a technogenic process of storing images and it would have anticipated Edison's kinetoscope. The only obvious difference lay in the represented narratives: while Edison naturally made his money with quintessentially American boxing matches, the peep show cabinets stored in the Salzburg museum showed scenes from the holy story with Joseph, Mary, Jesus and the disciples. Loyalty to the Counter-Reformation and Jesuit propaganda could hardly be carried any further.

2.2.3.5 Jesuit Churches

But the peep show cabinets displayed at fairs simultaneously also served as a model for what was supposed to be crucial for the elite public, or for court and city, as it was called at that time: the peep show theater. Unified perspectives, changeable backdrops, theater

81

lighting as in the small transportable peep show cabinets – everything that had previously never existed in the theater. I name as the most necessary key words only the sun lighting of the Theater of Dionysus in Athens and the architecture of medieval mystery plays, which also took place outdoors and (like Dante's *Divine Comedy*) played on three stages or showed parallel narratives at the same time: above in heaven, below in hell and in between them on earth. Anglicists know better than I do how many elements of Shakespearean theater from 1600 were still in line with this old European theater, which was allegorical but for that reason also had no illusions. But after St. Ignatius caught a glimpse of total hell through the art of meditation in his tiny cell in the cloister, and the Counter-Reformation of optical deception was under way, such simplicity was no longer enough. The same wave of innovation influenced both the interior architecture of churches and theaters. Namely, strictly following Lacan, it involved projecting pictorial linear perspective back into architecture again. What appeared to spectators and users in three-dimensional space should obey the same principles that Brunelleschi and Alberti had established for the two-dimensional space of the painting. And because the theater was employed as a church for the propaganda of faith, the church conversely also became a theater and the sacred building was seized by this perspectivization of architecture.

One of the most important innovations of baroque architecture was the well-known use of *trompe l'oeil* paintings, which provided easy transitions between buildings and paintings, three-dimensionality and two-dimensionality; for the deceived eyes the ceiling paintings of heaven, with all of their saints, became an integral part of the church and the perception of perspective. One becomes aware – and I will return to this later in my discussion of the image technologies of the Romantic period – of the almost cinematic sex appeal of the saints, who, like Bernini's statue of St. Theresa of Avila, seem to be engaged in a love act with the angel of their hallucinations even though and precisely because they are part of the church architecture.

As a matter of fact there is evidence – even if it is only later, namely Romantic – that this new image technology of church space was connected to the *lanterna magica*. A story by E.T.A. Hoffmann with the distinctive title *The Jesuit Church in Glogau*, which just as distinctively comes from the first part of his *Nachtstücke* (Night Pieces), can be traced back to his own experiences in the Silesian city of Glogau, where there actually was a large Jesuit church. Hoffmann begins quite matter-of-factly:

82

Jesuit cloisters, colleges, and churches are built in that Italian style which is based on antique models, and prefers splendour and elegance to ecclesiastical solemnity. (Hoffmann, 1952, p. 70)

A Jesuit professor in the novella rationalizes this deviation from the Gothic, i.e. the supernatural, with the following explanation:

[T]he higher [supernatural] kingdom should indeed be recognized in this world. But this recognition can be stimulated by symbols of joy, such as life offers, as does the spirit itself when it descends from that other kingdom into our earthly existence. Our home is there above; but so long as we dwell here, our kingdom is also of this world. (Hoffman, 1952, pp. 70–1)

The main character of the novella clarifies what is politically and technologically meant by this very ambiguous explanation that the Jesuit kingdom is also of this world. The hero is a painter from 1795 who is hired by the Jesuits in Glogau to restore the church. Because real marble is too expensive for the Jesuits, he is supposed to paint the columns of the church in such a way that it simulates the appearance of marble (Hoffmann, 1952, p. 71). One night, the sleepless narrator walks by the church – Hoffmann is after all narrating one of his night stories (*Nachtstücke*) – to eavesdrop on the highly dramatic scene between a *lanterna magica*, a *trompe l'oeil* picture, and a technically remote-controlled painter:

[A]s I passed the Jesuit church, I was struck by a dazzling light streaming through one window. The small side-door was ajar, and when I walked in I saw a wax taper burning before a tall niche. On drawing nearer I perceived that a packthread net had been stretched in front of this niche, and that a dark figure was climbing up and down a ladder, apparently engaged in making a design on the niche wall. It was Berthold, and he was carefully tracing the shadow of the net in black paint. On a tall easel beside the ladder stood the drawing of an altar. I was astonished at the ingenuity of the idea. If you have the slightest acquaintance with the art of painting, you will know immediately what was the purpose of the net whose shadow Berthold was tracing on to the niche. He was going to paint an altar in apparent relief, and in order to get a correct enlargement of the small drawing he had, in the usual way, to put a net both over the sketch and over the surface on which the sketch was to be reproduced. But this surface was not flat; he was going to paint on a semicircular niche. His simple and ingenious contrivance, therefore, was the only possible way of obtaining a correspondence between the straight lines of the sketch and the curved criss-cross ones thrown by the net on the concave surface, and of ensuring accuracy in the architectural proportions which were to be reproduced in perspective. (Hoffmann, 1952, pp. 73–4)

Like the *lanterna magica*, therefore, the painter Berthold's arrangement is precisely a projection of light onto the foil or the background of darkness. The wax torch functions as a light source and the net of threads functions as a pattern, which, exactly like the grid in Dürer's instructions for perspectival painting, causes a geometrization of the pattern to be painted, namely the altar, while the semicircle of the niche functions as a projection surface. But the goal of the entire arrangement is avowedly to solve the geometric problem of how one plane, which is only the two-dimensional illustration of a three-dimensional body, can be depicted on a half-cylinder whose upper end is presumably shaped like a half-sphere – and the whole must be achieved by imaging or mapping without resorting to the non-geometric, namely arithmetic tricks of Cartesian geometry. The painter Berthold thus does not need to solve any quadratic equations, not to mention trigonometric equations, but rather, just as the lecture on Renaissance perspective painters describes, the hand of the painter very simply and automatically follows the preset lines provided by the projected shadows of the net. What emerges through this process, as the title of the novella says, is a Jesuit church and thus an altar that does not actually exist but whose architectonic three-dimensionality is only simulated through the deception of the very earthly eyes of the churchgoers. The inner hallucinations caused by the spiritual exercises of the order's founder, Ignatius Loyola, were reduced to the earthly realm of optical illusions for the lay public like the painter and the churchgoers he deceived. With reference to Berliner Bauakademie director Johann Albert Eytelwein's two-volume handbook on perspective (see Hoffmann, 1967, p. 796), Hoffmann's Berthold expressly formulates:

Because I've drawn this entablature correctly from the point of vision, I know that it will give the spectator the illusion of relief. (Hoffmann, 1952, p. 77)

Hoffmann had the best historical reasons for attributing the invention of a church *trompe l'oeil* technique to the Jesuits. As far as I know, the engineer-artists of the Renaissance adhered to the Euclidean right angle – not out of mere laziness, but rather from a love of mathematical economy. They thus always projected the effects of their linear perspectives or *camera obscuras* only onto plane surfaces. Baroque architecture, which dreaded the right angle like the devil dreads holy water, first raised the issue of how linear perspective could also be projected onto curved surfaces – the issue, in other words, of how architecture itself could be included in the illusionistic game of new paintings. The theoretical solution of this problem appeared in 1693 in the form of a book with the very distinctive title, *De perspectiva pictorum atque architectorum*, about the perspective of painters *and* architects. Its author was, just as distinctively, a Jesuit priest named Andrea Pozzo. And the whole reason for the new *trompe l'oeil* effect became apparent at the very latest when Pozzo added a ceiling painting to the church of the founder of his order, Sant' Ignazio in Rome, which Jacob Burckhardt could not help from celebrating or castigating as a "playground of unscrupulousness." For this painting not only extended the architecture of the church into the illusionary heights of heaven, but it also subordinated all of its columns and saints, ledges and clouds to a monstrously distorting linear perspective that depended more on the elliptical curvature of the vaults than the subaltern and earthly vantage point of the devout. With his half-cylindrical curvature Berthold proves himself to be a direct pupil of Pozzo. Why the romantic narrator, that is, Hoffmann himself, also serves his apprenticeship under Berthold's Jesuits remains an open question that we must defer. So much for the Jesuits' use of image technology within the space of the church itself.

2.2.3.6 Jesuit Theater

The Jesuits also attempted to reform the profane or worldly theater using their new media technology: there was Jesuit theater. Like Kircher's optical devices or Hoffmann's church painter, Loyola's hallucinations were supposed to be mediated to a lay public without the strain of meditation. The most famous example from German literature is the *Cenodoxus*, which was written in 1602 by a Jesuit theology professor named Jakob Bidermann, who later was to become assistant to the Superior General in Rome. *Cenodoxus* presents a Parisian scholar who merely feigns a belief in Christianity because of his sheer pride of knowledge and is therefore carried off to the flames of hell by an entire horde of demons like the legendary Faust prior to Goethe's critical intervention. It is not difficult to read the drama as the victorious campaign of images and hallucinations against a form of knowledge rooted in the letterpress. But this is why Bidermann was not content to put placeless and timeless allegories on a neutral stage. Rather, his Jesuit stage presented concrete interior spaces that were divided towards the side of the spectators, and at the end of these spaces was a trapdoor in the floor of the stage that also guaranteed the direct entrance to hell (Bidermann 1965, p. 163). In the finale, when countless demons stormed out of this gate with pitch and sulfur and into the scholar's study in order to fetch Cenodoxus's body, his students were also present. For these students, who represented the actual spectators themselves, the Jesuits' spiritual exercise thus became a theatrical and sensorial certainty.

It should also be emphasized that the Jesuit stage and the related Italian opera stage stood as models for all of the new baroque theaters; in other words, the theaters of absolutism. For the first time, stages became peep shows, as we still know them today. These theaters confront us spectators with illusions that are more or less successful depending on the skill of the artisans and the financial framework of the theater technology. Perspective, which in Brunelleschi's time still urgently needed experimental proof, and which in Dürer's time still needed the support of scientific and technical grid constructions, suddenly became part of everyday life, or really night life, as a peep show stage. In the Silesian baroque, this golden age of German literature around 1670, all of the illusions created using the *lanterna magica* by Jesuits and traveling entertainers could finally be staged dramatically, which means no longer scaled down: there were painted perspectival backdrops that seamlessly extended the interior architectonically constructed stage setting and that were also interchangeable while a performance was in progress; there were costumed actors who transformed into other people on the open stage through a change of clothing (as in Gryphius's *Leo Armenius* or Lohenstein's *Agrippina*). It was thus practically tested how much could be asked of the spectators without breaking the illusion through effects, transformations, and conjuring tricks. The popular and thus quite underground history of such theatrical techniques leads directly to the nineteenth century, when the peep show even learned proto-cinematic turns and rotations from the *lanterna magica*: the *Unterklassentheater* in Josephstadt in Vienna introduced the first moving stage in 1842 in order to convey to the spectators the illusion of a real boat ride along the Danube river; in 1896, Karl Lautenschläger finally installed the first revolving stage in the Residenztheater in Munich, on which Gurnemanz and Parsifal were then able to hold their famous conversation: as we know, both of Wagner's opera characters were on their way to Gralsburg when, according to the director's instructions, "the scene changes imperceptibly." The pure fool Parsifal, this prototype of all the pure fools that Wagner's musical drama strategically had in sight as its spectators, did not notice and instead attributed the illusion to his own self-perception: "I scarcely move, yet swiftly seem to run." Whereupon the fatherly spirit or commentator Gurnemanz only needed to explain: "My son, thou seest here space and time are one." (Wagner, 1938, p. 445)

But we will return in due course to Wagner's feat of going from theatrical sleights of hand to an entire metaphysics and thereby motion pictures. Here, where we are concerned with the baroque and absolutism, the technology of illusions must be limited in a crucial way that will be important for the history of film.

The theater scene that turned into the image technology of the Counter-Reformation could manipulate and simulate nearly everything except its light source. For the first time in history – and at exactly the same time as absolutist castles – a closed theater whose narratives mainly took place in interior spaces and whose performances preferably took place in the evening needed artificial light. Richard Alewyn's *Barockes Welttheater* (Baroque World Theater) showed once and for all the price of no longer going to bed at the same time as the chickens. In 1650, all that was available to manage this time delay, which remains fundamental for all entertainment media ever since, were wax candles and torches – light sources, therefore, that almost gave off more heat than light and, based on Shannon, also served as involuntary sources of noise: they produced a sensory noise known as smoke and stench, which probably undermined the theatrical illusion except in scenes depicting hell, such as with Bidermann's. Unlike the backdrops, actors, and costumes, however, the hundreds of candles that were used on the stage as well as in the auditorium could not be changed during a performance. The dramatic but completely forgotten result of this limitation was the fact that none of the famous dramas by Corneille or Racine number more than 3,000 alexandrines. Hermeneutic literary studies has actually found the most beautiful and completely textually immanent explanations for this, but they are all worth little, because this aesthetic restriction follows immediately from a technical restriction: namely, the burning time of wax candles. In other words, Racine's Phaedra (who is a granddaughter but in contrast to Greek heroes only the granddaughter of the "holy sun") must die not because the flame of her incestuous love for her stepson burns so black, as she complains, but rather because after two hours the smoky candles in the Paris theaters had burned out.

But apart from the problem of light sources, absolutist theater was a mobilization of the theater machine for the actual deception of spectators. This was even true for Racine, whose heroines principally fall in love with an optically beaming vision (Barthes, 1964, pp. 19–20). After kings and princes systematically transformed themselves into optical illusions through their mirrored rooms, festival parades, and fireworks – before which their subjects could only look on in awe – their stages could not be left behind. Absolutism introduced a politics of images, and the effects of images in politics today – the portraits of heads in newspapers and television interviews on Sunday evening, this old absolutist monopoly on light – seem almost weak in comparison.

Within this historical context Bertolt Brecht's attack against what he called Aristotelian theater, which invites spectators to identify with and recognize characters, also appears comparably weak. In the materialistic light of media history, which neither Marx nor Lenin wrote, Brecht's attacks against traditional theater in his *Short Organon for the Theater* are unfortunately simply misdirected: the moments weakened by the deception of spectators do not all come from the open theater under the Greek sun or from Aristotle's *Poetics*, but rather from the peep show theater of the Counter-Reformation and its image war. Brecht the propagandist simply misjudges his historical predecessors. It would even be entirely appropriate to show that his extolled antidote to the theater of illusions – the allegedly scientifically proven alienation effect – practically coincides with a technologically modernized optics. It was the motion picture, whose arrival finally made the peep show stage of illusions, which had prevailed since the baroque, once and for all look old, and it developed into the great rival of all theaters.

That is all I will say about theater studies for the time being, but I will return to this topic when the introduction of new technical light sources comes up. To stick to the chronology, the question arises as to how the new optical media changed literature via theater. The impact on books is important because they are the only things in a position to provide evidence of the real bodily effects of media at times when there were no reader surveys or experimental psychological studies. Without knowing how the Romantic period attempted to compete with the precision of images, the preconditions and content of motion pictures will remain in the dark.

2.3 Enlightenment and Image War

We thus come to the eighteenth century, which was called the Enlightenment in English and the *siècle des lumières* in French, the century
of lights, as if it wanted to celebrate the recent optical inventions
that had been made. In this century, a technique of writing emerged,
or more precisely a technique of description, that, if you will, made
texts compatible with screenplays for the first time. When read correctly, the letters on the page enabled something to be seen – as if St.
Ignatius's solitary Bible readings had been applied on a massive scale.
  At the beginning of these developments and at the beginning of
this century stood, as the Germanist August Langen recognized long
ago, the so-called *Rahmenschau*. In Langen's terminology, this simply
means that from this time on poems also recorded nature using
techniques that had already been introduced to painting and theater
through the *camera obscura*, the *lanterna magica*, and the peep show.

2.3.1 Brockes

We will take the most famous and eloquent example: a poem that
Hamburg civil servant Barthold Hinrich Brockes published in his
nine-volume collection of physical and moral poems beautifully entitled *Earthly Delight in God* (1721–48), which is even more beautifully entitled *Bewährtes Mittel für die Augen* (Proven Remedy for
the Eyes):

When we stand in a beautiful landscape surrounded by loveliness,
Stirred by creation, more attentive than ever before,
To observe and actually appreciate its adornments,
To feel reasonable desires once more; so we find that our eyes
(Almost blinded and made clumsy by force of habit)
Are not properly suited to see the fine patterns, the colors, the harmony
    and splendour,
In which they are too scattered.
It appears as if our thoughts are also scattered, like our eyes,
And this is the sad reason why we are not able to delight in the world
Or glorify God through His creation with more enthusiasm.
With the bright light in our seeing crystals we are overwhelmed by too
    many patterns at once and actually from all sides.
Instead, our reason should endeavor to unify them,
Observe them one after the other, admire them,
And take pleasure in doing so: thus the spirit returns
Just as light and sight suddenly return,

Without bringing adornment and neatness to the body, as is necessary,
Without bringing out the delight, gratitude and thanks in us.
To dam the source of disinclination and ingratitude in us, the true
   fountain of unhappiness,
And still to do something, to look at the human species,
And to make us able to see: I suggest applying a remedy
Which was inspired by the observations that affected me as I was
   recently walking in the field,
And which is not difficult at all to use [. . .]
In a flat, open field, in which you go walking,
And in which you see nothing other than the field and the sky,
I want to point out to you a thousand different beauties instead of a
   landscape.
One simply has to fold up one of our hands
And hold it before the eye in the form of a perspective;
Through the small opening, the visualized objects
That are part of the general landscape are transformed into their own
   landscape,
From which a nice depiction could be drawn or painted,
If one is able to paint. It is only necessary to turn the hand slightly;
A new and entirely different beauty will immediately be seen.
The reason that beauty is so varied for us
Can be explained: the number of patterns in the eyes is so vast
That we are unable to differentiate between them properly,
And we are deterred, and the rays that strike the nerves of the face
And convey the figures of bodies onto the reflective crystals of the eyes,
No longer become clearer as our spirit notices them more sharply;
The gentle darkness of the small shadows formed by the cupped hand
Strengthens the eye, and consequently the spirit is sent,
Directed towards things and their details with greater attention and
   precision,
To consider the beauty within them with greater emphasis.
It is not contradictory, and it remains a consistent truth
That what Newton wrote about the sense of sight,
Is still hidden for many people, as it also was in the past:
Seeing is an art, like writing or reading.
In order to see properly, reason must be used,
Just as other senses are often used more for all other ends. (quoted in
   Busch, 1995, p. 113)

Since Brockes, poets also arm their eyes – not with an actual small
telescope aimed at earthly targets (which is what the word perspective
implied at that time), but rather with its literally manual replacement.
With an absolutely current invocation of Newton's *Optics*, sight
becomes an art that would not exist at all without the artificial aid

of devices like the peep show cabinet. But the new art of seeing does not merely position itself at the level of much older arts like writing and reading; it also brings its own innovation to these two sides of literature.

After the nymphs and gods left Greece, poetry was a tedious business: for two millennia it was necessary to rewrite a theme using set rhetorical phrases as if nymphs and gods still existed. Reading accordingly meant copying the arts of old texts and practicing with one's own quill for future texts. These rhetorical conditions, which dominated all schools and universities with the exception of the Jesuits' spiritual exercises, had nothing to do with writing and reading, but rather (as a commentary on Aristotle in late antiquity already noted) they only dealt with tactical references between senders and receivers, speakers and hearers, writers and readers. Imitating or surpassing an author – that was entirely different from the situation pertaining to writing. Impressing or even overwhelming readers with rhetorical effects and arguments until they no longer recognized truth and falsity – also pure cunning.

It is very different with Brockes, whose *Physical Poems* would not have existed without the lyrical imagination, representation, and even production of natural things. Poetry thus becomes – in the rigorous terms of Heidegger's history of being – an activity performed by subjects on objects, by writers on the beauty of nature, which practically cuts out the address because it relates to nothing but objects. In *Proven Remedy for the Eyes* poetry readers are only addressed when the writer presents them with his natural optics as a model. He and they do not restrict themselves to the many words about things that have already been made, stored, printed and could have even become mathematical since Pascal or Kircher's combinatorics. Instead, writer and reader adhere to the extra-linguistic, perspectival actuality of objects as such. This imposes a new form of writing on them, which is called (three centuries after Brunelleschi) perspectival. In the future, texts must be written so that it is possible for readers to reconstruct a view of the object written about in the poem even when they have never viewed the object themselves. Indeed, not only must they be able to reconstruct a view of the absent object, but they must also be able to reconstruct the view or perspective taken by the equally absent author with regard to the object. Heidegger rightly calls such image technology "the age of the world picture" or world pictures, but only in order to be free of it. All that remains in this lecture, which can only support Heidegger's high aims through analyses, is to note that the suppressed rhetoric avenged itself terribly on this forgotten writer-reader relationship, abandoned in favor of the new collective object relationship – in the form of ghosts, who will haunt us directly.

For the moment, however, I would still like to add something about how Brockes' manual peep show manufactures the object relationship. Brockes himself says clearly enough that his lyrical process is an artwork, an intentional interruption of established seeing habits, which normally ensure that our divinely willed and divinely oriented earthly delights do not sufficiently register the many details of this earthly reality. In this respect, a hand cupped in front of an eye to form a telescope creates an alienation effect, long before Brecht, simply because a poet with his armed eye declares war against the unarmed. One need only exchange the meek Brockes with his more aggressive writer colleagues in Ireland and France to understand the keenness of this scientifically enlightened technology of vision. For example, when Swift's Gulliver comes to the land of the giants – which, as we know, lies only a little north of San Francisco – he does not need any perspective or microscope at all, because the giants already appear ten times larger to his human eyes. However, the sad result of this microscopy built into nature is that when the giant queen's most beautiful 16-year-old lady-in-waiting allows Gulliver, like a pet, to climb around her breasts or ride on one of her nipples, he is not aroused at all: he sees not only the nipples themselves, but also the individual skin pores all around them as enormous warts.

> The handsomest among these maids of honor, a pleasant frolicsome girl of sixteen, would sometimes set me astride upon one of her nipples, with many other tricks, wherein the reader will excuse me for not being over particular. But I was so much displeased that I entreated Glumdalclitch to contrive some excuse for not seeing that young lady any more. (Swift, 1963, pp. 137–8)

It is precisely such warts that Denis Diderot, the first literary theorist of realism, must use as an excuse to make characters that the writer invented out of nothing nonetheless appear perfectly believable and true-to-life for readers. Diderot argues that when the reader finds a mention of a wart on the face of the literary heroine, he cannot avoid thinking that the writer could not have invented inconspicuous details like warts and must have described them according to so-called life instead. The ugliness that serves as realistic content in Swift's text is thus magnified or abstracted by Diderot into the textual form of realism (Jauss, 1969, p. 237).

92

To summarize briefly: since the time of Diderot and his translator Lessing, literature itself trades under the name of a quasi-optical
medium. Lessing's treatise on *Laokoon* systematically compared
poetry to painting and concluded with the following imperative concerning the effects of poetry: the poet should make "his subject so
palpable to us that we become more conscious of the subject than of
his words" (Lessing, 1984, p. 75). Endowed with this mission, but
still nevertheless without any guarantee of its automatic realization
on the part of the reader, literature abandoned the realm of letters,
which had been well-defined since the time of Gutenberg and Luther,
and it became a virtual light that fell on the objects it described – in
other words, an enlightenment.

In the field of optical effects, however, another power was already
established after the Jesuits' Counter-Reformation. With its new
assignment – to place objects before the eyes of a public that needed
to be enlightened – literature necessarily assumed a combat position.
It became, as Hegel's *Phenomenology of Spirit* so strategically and
elegantly formulated, "a battle between the Enlightenment and superstition." And, as Jean Starobinski added, this battle was played out
between light and darkness. In this respect, the French Revolution
was described as an optical medium, as you can read in Starobinski's
1789: The Emblems of Reason. However, Starobinski failed to look
beyond the metaphors and to follow the battle between light and
darkness taking place right inside his actual concrete devices. That
remains to be done, and it will finally take us back to the history of
the *lanterna magica* and allow us to understand the reason why the
literature of the Romantic period itself became a *lanterna magica*.

2.3.2 Phenomenology from Lambert to Hegel

The battle between the Enlightenment and superstition simply meant
wresting optical media away from the Counter-Reformation and
giving them a better purpose. Brockes already employed the *Rahmenschau* not for the sake of optical illusions, but to provide his readers
with a scientifically enlightened way of seeing. The science of optics
thus changed its function: from Aristotle to Huygens or Newton it
had virtually been a cosmic science, which benefited astronomy in
particular. In the eighteenth century, on the other hand, optics surrendered to the Cartesian subject that it had produced itself through
linear perspective, *camera obscura*, and *lanterna magica*. It no longer
simply asked how rays of light travel from the world into the eye
after all possible reflections and refractions, but rather it posed the

new question of how the optical actuality of the world can be reconstructed from the data available to the eye as sensations (to John Locke's definition of the word). If one could namely explain the differences that exist between the objective world and subjective actuality, the priests' deception that exploited precisely this difference would be completely exposed.

The first philosopher to formulate this question was the Alsatian Johann Heinrich Lambert. Lambert came to Berlin in 1764 to obtain a position at the Royal Academy of Science. At first, Frederick the Great was horrified by Lambert's primitive eating and drinking habits, which valued pasta and overly sweetened wines above all else, until proof of Lambert's brilliant knowledge of mathematics reassured him. It was this knowledge that first made a philosophical-mathematical theory of appearances possible.

Lambert had already submitted a paper in 1759 with the characteristic title *Freye Perspektive* (Free Perspective). "Free" was supposed to imply that the art of perspective taught to painters by Brunelleschi and Alberti had been very unfree. With fictional images, which removed the technical mediation of the *camera obscura*, the painter first had to abandon his inherited two-dimensional medium and think himself into the strange technology of architecture. According to Alberti's instructions, the proportions, lines, and vanishing points would only be correct when the painter was able to draw a ground plan and a side elevation of the space that his painting was then supposed to shorten using perspective. Lambert's free perspective wanted to relieve painters of precisely this toil. It could only do this, however, because the mathematician from Mühlhausen understood the greater mathematician from the neighboring city of Basel. Leonhard Euler succeeded in purifying four trigonometrical functions from all table work or empiricism. Because he standardized the radius of a circle as one unit, within whose sine and cosine, tangents and cotangents develop, table values became real mathematical functions that depend on one single, real variable (or, since Euler, they can also depend on complex variables). In other words, for the first time transcendental functions could be calculated elegantly and in general terms, which was previously only possible with algebraic and therefore polynomial equations.

Lambert applied Euler's mathematical trick to painting. He no longer determined perspectival geometry as relationships or proportions between lines, like Euclid or Pythagoras, but rather as transcendental functions of an angle of vision. Wonderful, I hear you say, but how were mathematically untrained painters supposed to know these functions? The answer was quite simple: Lambert removed this burden from their shoulders by spoon-feeding them the calculations. Lambert, who was brilliant at inventing new measuring instruments, invented a ruler that would have simply appalled Euclid or Dürer. In addition to the venerable linear scale, it also contained four other gradations that made all four trigonometrical functions – at least in the range from –90 to 90 degrees – available at the same time as the linear function. A painter who had obtained Lambert's ruler could cast an eye over the open landscape, slide the ruler over the drawing, and an image emerged in perspective entirely without a ground plan or a side elevation. As an amateur art historian, I unfortunately do not know if real or even famous painters employed Lambert's *Freye Perspektive* other than the Berlin artist Hummel, but hopefully its basic principle already clarifies how modern trigonometrical functions are capable of mathematically recreating visual appearance, which means the perspective of a seeing subject.

Lambert also employed the same trigonometrical trick in a branch of optical science that he himself invented: photometry. Before Lambert, nobody would have known how to say how bright or how dark what we see actually is. Photography, which is nothing but automatic Lambertian photometry, conversely made sure that we no longer even know about this prior lack of knowledge. If the light from a light source, whose radiated energy is hopefully constant, falls on a curved surface of the same diffuse color, as in Hoffmann's Jesuit church, then a literal appearance or reflection enters the eye, which depends not only on the brightness of the light itself, but also on the angle that the direction of the light forms with the local perpendicular on the surface. $l = k * cos \ (a)$ (Watt, 1989, p. 48).

Both examples, free perspective and Lambert's cosine law (which, incidentally, he himself still wrote up as sine law), were only supposed to show how Euler's and Lambert's new trigonometry made it possible to calculate a subjective appearance. All that remained, therefore, was to build this mathematics of the subject into an entire philosophy. Lambert did precisely that in his *Neues Organon* (New Organon), which was published the year he traveled to Berlin and already revealed the formulation of this new question in its subtitle: *Gedanken über die Erforschung and Bezeichnung des Wahren und dessen Unterscheidung vom Irrtum und Schein* (Thoughts on the Investigation and Designation of Truth and its Distinction from Error and Appearance). You know that all philosophies have always known that there is a gaping abyss between truth and error, which only they were supposed to bridge. The difference between truth and appearance is less venerable, but above all less clear, particularly when one realizes, as Lambert himself first explained, that with every perspective or diffuse reflection a visual appearance necessarily emerges. Lambert's *New Organon* (in contrast to Aristotle's good old one) also makes the point that appearance chiefly arises and is moreover chiefly researched in optics. To generalize his concept for all five senses, however, Lambert coins a new word that would go on to make philosophical careers: he establishes a doctrine of appearance in general, that is, a "phenomenology." As you can imagine, this science is concerned with the struggle between the Enlightenment and superstition or, in Lambert's words, "with the means of avoiding illusions and penetrating appearances in order to get to the truth" (Lambert, 1990, p. 645). For this reason, the first "examples" that Lambert uses to "explain" his program are taken from his own optics: "The color of a body depends on the light illuminating it. At night, scarlet and black can hardly be differentiated. By the light of a lamp, blue and green are practically the same. The sun affects not only the brightness of a wall, but also its color. The question is thus by which light does a body show its true color?" (Lambert, 1990, p. 674)

To begin with, therefore, phenomenology is the science that investigates the objective nature of things starting with the subject and particularly the subject's visual appearances, and in this respect it does away with all sensory deception or even priests. However, because Lambert deals not only with thought and recognition, but also with the description of this thought, the formulation of the question of phenomenology can also be reversed. As a brilliant mathematician, Lambert already knows that there would be no thought without signs and no calculation without symbols for numbers and operators. For this reason, his phenomenology also ends with a "sixth chapter" that does not undo appearance but rather constructs it: "In its general scope, phenomenology is a transcendental optics [. . .] as it determines appearance from truth and, in turn, truth from appearance" (Lambert, 1990, p. 824). The adjective "transcendental" clearly does not convey its traditional meaning, which denotes God's place outside the world. On the contrary, it is a mathematical term that Lambert takes from his transcendent trigonometrical functions and imports into philosophy. All that is left for Kant, Lambert's younger friend and pupil, is to go back and generalize this "transcendental optics" as something that is now called transcendental philosophy. In other words, which are unfortunately seldom used by philosophers: German idealism also emerged from the history of optical media. Hegel thus has good reason to treat what he calls the struggle of the

Enlightenment with superstition in a book that – both out of loyalty and disloyalty with respect to Lambert – is entitled *Phenomenology of the Spirit*. At least the early Hegel did not want to confront his readers immediately with God, but rather to construct the absolute as the last perspectival vanishing point after the abolition of all subjective appearances. This is all quite wonderful, but it will not be discussed any further in this lecture because transcendental optics forfeits every mathematical ambiguity as it passes from Lambert to Kant to Hegel.

On the other hand, the final chapter of the *New Organon*, entitled "On the Drawing of Appearance," provides a virtual summary of everything these lectures have dealt with so far. It addresses linear perspective and painting, linear perspective and theater architecture, and finally even linear perspective and the "art of poetry" (Lambert, 1990, pp. 824–8). As if to generalize Brockes' opthalmological advice or Diderot's wart, Lambert claims that the "art of poetry" consists of the art "of painting things for us according to their appearance and evoking the same sensations through these visions that the things themselves would evoke if we saw them from the same viewpoint as the poet's, which he transfers to our thoughts, so to speak. [. . .] [The poet] paints the aspect of the thing that he himself already entirely imagines from his assumed viewpoint, with all of the insights and desires that these impressions inspire in him and that they are supposed to inspire in his readers" (Lambert, 1990, p. 828).

The battle between light and darkness brought all the illumination effects that were formerly available only to aristocratic powers like royal courts or religious powers like the Jesuits into the simple, quotidian night of middle-class people. Artificial lights that shone until midnight were installed everywhere. Secret societies like the Freemasons, who as I mentioned before, could only have come into being after workers' secrets were revealed through print technology, organized their rituals partly in the dark of the night and partly in the illuminated night, and in this twilight it was never possible to discern whether the secret leaders were doing this to follow the goals of the Enlightenment or the Counter-Reformation. In the eighteenth century, therefore, secret associations represented the common ground upon which Catholics, free-thinkers, Jesuits, and enlighteners were able to wage their war. They also represented the ground, however, upon which the optical effects of Jesuit propaganda gradually turned into commercial (to avoid using the word fraudulent) practices.

The literature of the Enlightenment thus paid for the fact that it invented a concept of what objects are worthy of being represented, but thereby lost all strategic concept of their target. Con-artists who employed optical media to manufacture completely illusionary objects, but who did it before an audience they had strategically and exactly calculated, filled only one of the gaps left or rather provided by the system.

2.3.3 Ghost-Seers

The *lanterna magica* thus stopped projecting the Stations of the Cross or the sufferings of the damned in hell, as it once did for the baroque Athanasius Kircher. Instead, it became an all-purpose weapon used by adventurers like Count Cagliostro or Casanova, the seduction artist. The second and third books of Casanova's *Memoirs* recount how he played a magician at night in Cesena only to cheat an idiot who believed in the hidden treasures of ghosts out of 500 sequins and also naturally to seduce this fool's daughter. Casanova succeeded at this second project quite easily, as usual, but when he attempted to act like a magician a storm interfered. Its lightning frightened the magician himself far more than his artworks frightened the deceived spectators . . .

Even more dramatic than Casanova's unique experiment is the entire life of a con-artist whose surname already proclaimed his true intentions: the man was simply named Schröpfer.[4] Allow me to cite Schröpfer's swindle directly from Zglinicki's amusing *Der Weg des Films*, because in this story every word counts:

> Schröpfer owned a coffee house on the Klostergasse in Leipzig, which enjoyed a certain popularity, especially because there was an excellent punch there. The owner of the establishment understood masterfully how to surround his guests with the halo of a great magician! He discretely made it known that he was "in the possession of the true secrets of the Freemasons." Many sought this true secret at that time, for they were convinced that unlimited power would then be bestowed on them. There were also such doubters and seekers among the regular guests of the coffee house on the Klostergasse in Leipzig, some of whom were distinguished and held high offices. People of this sort made quite an impression on Schröpfer. He approached them carefully to question them and make them bend to his wishes. He never resorted to empty speech, however, and instead produced "facts." These "facts" consisted in so-called "necromantic" (today we would say spiritualist) sittings, which distinguished him especially through the

---

[4] The German word *Schröpfer* refers to someone who rips off other people.

use of magic lantern effects and sophisticated theatrical make-up. Even those who possessed the most level-headed and critical natures were taken in. The stage for these myserious performances was the billiard room of the guest house of the new "Grosskophta." Schröpfer had it built especially for this purpose. All of the spectators first received a large dose of punch. This not only increased his profits, but also his guests' psychological willingness to see miracles. After the punch had fogged their brains enough, Schröpfer moved to the billiard room with his guests and began the performance. To ensure that meddlesome and distrustful spectators would not see through his game, Schröpfer made various preparations and established strict rules of behavior. No one was permitted to stir from his place, and only a small lamp burned in the dark room, whose flickering light intensified the gloomy atmosphere. Incense, a skull, and other mystical utensils completed the "magician's" furnishings. Then the conjuration began: accompanied by dreadful rumbling, a preloaded ghost appeared constantly fluttering back and forth above a kind of altar. It became brighter and then darker again. Of course it was a typical *lanterna magica* image projected backwards against the rising smoke. Electrified machines ensured that the spectators received invisible and mysterious electric shocks at the high point of the manifestations, assistants imitated otherworldly voices, and much else. In the beginning, Schröpfer's bogus machinations were limited to the area around Leipzig, but soon he expanded his area of operations and undertook trips to Berlin, Frankfurt am Main and Dresden. Thanks to his skillful scheming and excellent relationships he soon succeeded in gathering the most prestigious and affluent circles around him. Schröpfer eagerly began to advertise something new: he wanted to found his own masonic lodge, and he actually received money at once from all sides. Sponsors and friends trustingly granted him enormous credit. In return, Schröpfer promised them – naming is destiny – high annuities from a fortune in the millions, which was allegedly deposited in a bank in Frankfurt. The magician was able to stave off his patient believers for a long time with all sorts of excuses. Finally – on the evening of October 7, 1774 – he promised to show all of his friends something from his fortune. Spirits were high in the Klostergasse. Everyone showed up, even the serious and wealthy gentlemen with the impressive titles. The punch was better than ever before. People jabbered on about "true wisdom" and "the eternal light." Schröpfer seemed to shine with the brilliance of his knowledge and fame – like a Pied Piper of human souls. He appeared to feel particularly well, and he was more amusing, energetic, spirited, and humorous than ever before. He behaved – as it was later recorded in the report – as if he wanted to go to a ball. It was not until midnight that he began to lose some of his spiritedness. He briefy withdrew and wrote a few letters. In the early morning hours he invited his guests to go with him to Rosental Park, where he would show them "a wonderful apparition." Still at dawn, Schröpfer went ahead of all the others. He asked his friends to wait a little, and they noticed how he continued along the way alone without looking around. He turned around the corner and was concealed by trees and bushes. His steps became silent, and then there was nothing more to be heard. The unnatural calm weighed over the park like a black cloth; the men looked at themselves, became restless – then there was a sound like the lash of a whip, which broke the silence. Schröpfer was found dead lying on the ground in the forest. A bullet in the mouth had ended his life. (Zglinicki, 1979, p. 67)

Compare this ending sometime with the ending of Sophocles' *Oedipus at Colonus* . . . Schröpfer's lovely story requires no interpretation, but only a few remarks and some added emphasis.

First, it makes clear what I have already said about the difference between the oral but technical traditions of medieval masons and the just as oral but occult traditions of eighteenth-century Freemasons. The letterpress, the *camera obscura*, and the *lanterna magica* automatized knowledge, and the drawback is that this results merely in assertions about knowledge whose only goal is to obscure the underlying technology, like a *lanterna magica*, using all means of intoxication.

Second, the story makes clear how the formerly aristocratic or religious night lights became bourgeois: coffee houses, which did not arise until 1683 (after the relief of the Turkish siege on Vienna and the subsequent capture of Turkish coffee supplies, which to Prince Eugene of Savoy's unexpected delight were incidentally mixed with hashish), profited from this new nocturnal brightness.

Third, the story makes clear how various drugs, from coffee to punch to the *lanterna magica* (not to mention hashish), all acted in combination to bathe such nights in a spiritualistic twilight. These drugs actually only needed to be used one after the other in literature to produce romantic literature, such as E.T.A. Hoffmann's *The Serapion Brethren* with its punch-drinking ritual or even his *Nachtstücke* (Night Pieces), which already incorporate the night into their title.

Fourth (and the most important for us), Schröpfer demonstrates that *lanterna magica* techniques made a leap forward in the eighteenth century: in the twilight of the artificial night it became possible for the first time to breathe technical life into projected spirits and ghosts. Even if this life was not yet created by the mechanism of film, but rather only the flickering of a curtain of smoke on which the *lanterna magica* projected its virtual images, Schröpfer's arrangement of magic lanterns and smoking pans shows very clearly how the desire for film technology was historically generated in the confusion of Enlightenment and superstition, inspiration and deception.

What Edison and the Lumières accomplished a century after Cagliosto and Schröpfer therefore fulfilled neither some timeless need nor some primal dream of humankind, which according to Zglinicki has supposedly been around from time immemorial; rather, it was a technical and thereby definitive answer to wishes that had been historically produced.

To provide evidence of this relay race from illusionists to engineers, there were actual magicians around 1800 who employed their *lanterna magica* for money and illusions, exactly like Schröpfer, but who at the same time also worked on scientific improvements of optical media technology. I will only mention one of these magicians: the Belgian Étienne Gaspard Robertson, who became known in the history of film for producing ghost projections that were more elegant or lifelike than Schröpfer's. Robertson was able to accomplish this by placing his *lanterna magica* on a wagon with large, lightweight and noiseless wheels, which could move around the room unnoticed like future film cameras. To magnify the illusion even more, Robertson particularly liked to appear in old cloisters, as if he wanted to recall the origins of the *lanterna magica* in the Counter-Reformation, and he filled the exhibition hall with skulls, bones, and memorial slabs, as if he already wanted to make an expressionistic silent film or orchestrate one of the mechanized ghost trains that appear at fairs (Zglinicki, 1979, p. 70). What film histories do not mention, however, is that in a completely different environment – namely, before the French Academy – this same Robertson also made scientific history on March 2, 1802. Instead of using electricity to dispense shocks from imaginary ghosts to superstitious spectators, like Schröpfer, or being frightened of the natural electricity of lightning, like Casanova, Robertson electrified two simple carbon rods using a voltaic battery, which had just been invented. He then gradually pushed the carbon rods closer together and thus triggered a spark between them, which blinded all of the stunned spectators for several seconds until the carbon burned and the light was gone. The carbon arc lamp – the first artificial light source that could compete with sunlight or lightning and that consequently became essential for photography and film – was invented.

2.3.3.1 Schiller

After this digression about Robertson as a bridge between the eighteenth and nineteenth centuries or illusionism and science, we return back to the problem of historically awakened wishes in general. The thesis was that in the battle between the Enlightenment and superstition, moving images were presented for the first time on a massive scale and thus became desiderata on a massive scale. Yet the more that magicians strove to fulfill the demand for moving images, the greater the strategic counter-wish to expose these images as mere illusions became. As the cases of Cagliosto, Casanova, and Schröpfer all sadly verify, this Enlightenment almost always succeeded, and in the cold early morning light one more suicide victim lay in the parks of Leipzig. That is why – and now comes the second thesis – the unfulfilled wish for moving images produced another medium, which could at least satisfy it in the realm of the imaginary for a period of time before the invention of film: romantic literature. Under this term I also include, in contrast to many Germanists, the so-called German or Weimar classic period.

When this is taken for granted, it is easy to prove the second thesis. Take, for example, the first German ghost novel of all, Schiller's uncompleted *The Ghost-Seer*, which was published in several thrilling instalments in the *Thalia*, Schiller's own newspaper, between 1787 and 1789. The hero of the novel is a German Protestant prince, who is only separated from the throne by a few senior relatives who are still alive. According to the arch-positivistic proof offered by a Germanist in 1903, this prince may actually have been a prince of Württemberg. It can also be proved that he believed in spiritualism or the practice of seeing ghosts, and after the death of Duke Carl Eugen, Schiller's own pedagogue or despot, he had well-founded prospects for the line of succession for a time. While this hero stops in Venice on the Cavaliers' tour, which was an absolute requisite for members of the nobility at that time, one of his bothersome relatives dies in a mysterious and unnatural way. The prince hears rumors in Venice that a group of mysterious people, including a so-called Armenian, are interested in eliminating these bothersome relatives and thus helping the prince himself achieve the princely line of succession. The first part of the novel, however, mostly deals with a ghost conjuration, which a nameless Sicilian is holding for the prince and his entourage in the hinterland of Venice on the beautiful Brenta.

We found in the middle of the room a large black circle, drawn with charcoal, the space within which was capable of containing us all very easily. The planks of the chamber floor next to the wall were taken up, all round the room, so that we stood, as it were, upon an island.

An altar, covered with black cloth, was placed in the center upon a carpet of red satin. A Chaldee Bible was laid open, together with a skull; and a silver crucifix was fastened upon the altar. Instead of candles some spirits of wine were burning in a silver vessel. A thick smoke of frankincense darkened the room, and almost extinguished the lights. The Sorcerer was undressed like ourselves, but bare-footed; about his bare neck he wore an amulet, suspended by a chain of human hair; round his middle was a white apron, marked with cabalistic characters and symbolical figures. He desired us to join hands, and to observe profound silence; above all, he ordered us not to ask the apparition any question. He desired the Englishman and myself, whom he seemed to mistrust the most, constantly to hold two naked swords crossways, an inch above his head, as long as the conjuration should last. We formed a half moon round him; the Russian officer placed himself close to the English lord, and was the nearest to the altar. The sorcerer stood upon the satin carpet with his face turned to the east. He sprinkled holy water in the direction of the four cardinal points of the compass, and bowed three times before the Bible. The *formula* of the conjuration, of which we did not understand a word, lasted for the space of seven or eight minutes; at the end of which he made a sign to those who stood close behind to seize him firmly by the hair. Amid the most violent convulsions he called the deceased three times by his name, and the third time he stretched forth his hand towards the crucifix. On a sudden we all felt, at the same instant, a stroke as of a flash of lightning, so powerful that it obliged us to quit each other's hands; a terrible thunder shook the house; the locks jarred; the doors creaked; the cover of the silver box fell down, and extinguished the light; and on the opposite wall, over the chimney-piece, appeared a human figure, in a bloody shirt, with the paleness of death on its countenance. "Who calls me?" said a hollow, hardly intelligible voice. (Schiller, 1904–5, II, pp. 248–9)

It later comes to light that the figure rudely chasing away the phantasmagorical figure of the spirit conjured by the Sicilian is none other than the Armenian. He has the Sicilian arrested by the Venetian police and forces him in jail to reveal to the prince all the technical tricks involved in producing magic. The readers of the novel then learn, along with the prince, that the ghost of the deceased was projected onto an artificial curtain of smoke using a *lanterna magica*, and the flash of lightning, which was felt by everyone present, was triggered by a hidden source of electrical energy – presumably a Leyden jar, as voltaic batteries did not yet exist. For those of you attending these lectures, however, this systematic enlightenment is hardly necessary:

you will have already recognized the similarities between the events organized by the Sicilian and those organized by Schröpfer, and in the fictional Venice you will also have recognized the historical-empirical Leipzig coffee house.

I begin the new year, the last one of this millennium, with the wish that it has begun well for all of you and that it will also continue to go well. Naturally, that does not mean that all of the good resolutions you would like to make must also be fulfilled. What will be fulfilled first is only my good resolution to begin the new year by continuing to the second major part of these lectures, which deals with optical media technologies. In today's lecture, the tales from the history of art and literature, with which I have sought to entertain you up until now, come to an abrupt end.

I have already emphasized that Schiller's novel was published in instalments, which were designed to produce more suspense for the reader. This is the reason why a magical Armenian suddenly emerged from the circle of ghost conjurers and put a stop to the Sicilian's deceitful game by whipping up a three-dimensional ghost instead of a two-dimensional one. As if to allegorize Hegel's battle between the Enlightenment and superstition, therefore, the Armenian hovers between the power of disillusionment and the power of creating even more remarkable illusions. On the one hand, by handing the Sicilian over to the Venetian police and then systematically interrogating him in prison, the Armenian provides the prince with a technically pure enlightenment that is above all the tricks that marked the conjuring of ghosts. In the second part of the novel, on the other hand, this apparent enlightenment proves to be a stratagem under whose protection the Armenian is able to go from exposing first-degree magic to producing second-degree magic without arousing any suspicions. It even turns out that the Sicilian was only one of the minions employed by the Armenian himself, and his intentionally transparent deception was supposed to set the stage for the actual deception. The novel is not about ghostly apparitions for their own sake, out of pure curiosity so to speak; rather, it is about a German and thus an enlightened and absolutist prince who is made to believe once again in apparitions. The perfect example of this optical belief is called Catholicism in the novel, and more specifically the Jesuit Order. Only a decade after most Central European states, including even the Vatican itself, suppressed the Jesuit Order, everything revolves around the machinations of an order that seeks to regain its power over German princedoms by either murdering Protestant heirs to the throne or converting them to the only true church. The early modern rule *cuius regio, eius religio*[5] was valid, at least pro forma, until the time immediately preceding Napoleon's dismantling of the old empire. All of the illusionary techniques that Schiller borrowed from his close knowledge of Schröpfer were accordingly transferred from the German swindler to the secret Catholic organization, and the theory of the Counter-Reformation presented here receives a literary support that can also be supplemented with a philosophical one without any difficulty. Kant's *Critique of Judgment* furnishes proof that for human powers of imagination (and actually without paradox) the most sublime thing is the imageless God – in the sense of Mosaic law – and conversely, every illustration of religion is already a governmental abuse of power that can only support the Counter-Reformation:

> [W]here the senses see nothing more before them, and the unmistakable and indelible idea of morality remains, it would be rather necessary to moderate the impetus of an unbounded imagination, to prevent it from rising to enthusiasm, than through fear of the powerlessness of these ideas to seek aid for them in images and childish ritual. Thus governments have willingly allowed religion to be abundantly provided with the latter accompaniments, and seeking thereby to relieve their subjects of trouble, they have also sought to deprive them of the faculty of extending their spiritual powers beyond the limits that are arbitrarily assigned to them and by means of which they can be the more easily treated as mere passive beings. (Kant, 1951, p. 115)

In Schiller's novel fragment, it is precisely this Counter-Reformation stratagem that triumphs: on the last published page the prince has just attended his first Catholic mass. To work such wonders of conversion, however, it was not enough to pull the wool over the eyes of enlightened aristocrats with *lanterna magica* gadgets like those of Schröpfer or the Sicilian. Quite apart from a few tricks with telescopes, the Armenian had to supply the prince above all with a woman, with whom a romantic like the prince could not help falling in love. This woman, a so-called Greek woman who naturally spoke German and was of the most noble German descent, thus appeared to her lover for the first time in disguise and devoutly Catholic in one of Venice's famous baroque churches surrounded by the holy paintings and ceiling frescoes of the Counter-Reformation. It is like

---

[5] A Latin phrase meaning "Whose realm, his religion," which refers to the compromise by which princes were allowed to determine the religion of their territories.

a stage production of love at first sight, on which romantic love by definition is based, and at first glance it appears to be a novelistic[6] relapse from Schröpfer to Tintoretto, from the *lanterna magica* to the traditional panel. But far from it! When the prince instructs his servants and escorts to try to track down the Greek woman, who disappeared again immediately after going to church, the medium that instilled this irresistible romantic love in the prince is finally mentioned. As we are all painfully aware in 1999, there are of course media technologies without love, but there is no love without media technologies. In the case of Schiller's princely ghost-seer, this eroticizing medium is by no means painting, but rather literature. The only reason why it was impossible for his escorts to track down the Greek woman in Venice is that the prince could not describe her at all due to his passion for writing literature. To cite Schiller's words:

> [U]nluckily, the description the Prince gave of her [the Greek woman] was not such as to make her recognizable by a third party. The passionate interest with which he had regarded her had hindered him from observing her minutely; for all the minor details, which other people would not have failed to notice, had escaped his observation; from his description, one would have sooner expected to find her prototype in the works of Ariosto or Tasso than on a Venetian island. (Schiller, 1904–5, II, p. 335)

I will come back later to the problem of wanted posters, passport controls, and forensics, which was theoretically as well as practically insoluble before the invention of photography. For the moment, however, it will suffice to ask why the prince could not see or describe the Greek woman he loved. The answer is in the text: he did not perceive the image of this woman at all, but rather he read it in the most famous verse romances by Tasso or Ariosto. His love is the kind of love found in romantic novels, and the Armenian's strategic art lay in replacing the transparent optical illusions of the *lanterna magica* with the opaque illusions of romantic literature. The explicit villain in the novel thus employs precisely the same medium as the novel's author. And because Schiller does not leave any doubt that his novel is one of the countless enlightened broadsheets against crypto-Catholic and above all Jesuit machinations, this parallel between the

---

[6] Here, Kittler employs the term *romantechnisch* (literally "novel-technical"), which is a play on the word *romantisch* (romantic).

novel's villain and the novel's author implies that romantic literature itself proposes to take the place of the techniques of illusion of the Counter-Reformation in the great image war. Schiller's ghost-seeing prince consequently proves that the claim Lessing put forward only as an imperative or objective – that readers were supposed to be more conscious of the author's ideas than the words written down and read – had in the meantime created "real" readers. I could discuss here the new pedagogical techniques used in reading and writing instruction around 1800, which as a historical novelty helped even small children learn to read silently. As for the goal of this training, it is sufficient to mention Hegel's famous dictum that only silent reading could serve as the pure foundation for the interiorization of every subject, and this interiorization would then give rise to pictures of the read material. If there should still be any doubt as to whether Schiller's fictional prince serves as a model or role model for all actual readers, it disappears when reading the sentence with which Schiller signalled to the readers of his newspaper that the novel would be temporarily ending, which also unfortunately turned out to be the final end of the novel fragment. Schiller wrote:

> To the reader who hoped to see ghosts here, I assure you that some are yet to come; but you yourself see that they would not be deployed for such an unbelieving person as the Prince still is at that time. (Schiller, 1904–5, II, p. 426)

"Here" in this passage clearly meant: here in this newspaper, here in my novel, here on these printed pages, of which hundreds or thousands of identical copies could be manufactured ever since Gutenberg's invention of the printing press. It is a fantastic claim that literature, as far as I see it, had never made either previously in old European times or later under technical conditions. In the classical-romantic epoch, however, this claim was self-evident for writers as well as for readers. Only a few years after Schiller's novel fragment, Leipzig scholar Johann Adam Bergk's book entitled *Die Kunst, Bücher zu lesen* (The Art of Reading Books) was also published in Jena. In its fight against religious superstition, this book was just as resolute as *The Ghost-Seer*, but it did make one exception: namely, if spirits can appear despite all religious skepticism, then they are only to be found in intellectually stimulating books (Bergk, 1799).

Such comments are not historical plays on words that simply swing back and forth between spirit in the sense of meaning and spirit in the sense of ghost. Rather, they describe a style of fictional

107

text that is also of the utmost importance, especially for film and television studies. To put it plainly: in contrast to certain colleagues in media studies, who first wrote about French novels before discovering French cinema and thus only see the task before them today as publishing one book after another about the theory and practice of literary adaptations – in contrast to such cheap modernizations of the philological craft, it is important to understand which historical forms of literature created the conditions that enabled their adaptation in the first place. Without such a concept, it remains inexplicable why certain novels by Alexandre Dumas, like *The Three Musketeers*, have been adapted for film hundreds of times, while old European literature, from Ovid's *Metamorphoses* to weighty baroque tomes, were simple non-starters for film. Even when the *Odyssey* is adapted in Hollywood, it is only adapted the way a nineteenth-century novelist would have retold it.

In her book, *The Haunted Screen*, Lotte Eisner precisely and correctly pointed out how early feature films returned to the themes and techniques of romantic literature (Eisner, 1973, p. 40), but she offered no explanations for this.

If I guess correctly, on the other hand, the difference between filmability and unfilmability, romantic and pre-romantic texts, lies entirely in the media image war that romantic literature won for a time against opponents like the church, which gradually surrendered. The winner in the battle between the Enlightenment and superstition, as in all of history, was the return of the same. In other words, the production of books, otherwise known as the Enlightenment, itself became an image technology.

It is even possible to go one step further and conclude from the visually hallucinatory ability that literature acquired around 1800 that a historically changed mode of perception had entered everyday life. As we know, after a preliminary shock Europeans and North Americans learned very quickly and easily how to decode film sequences. They realized that film edits did not represent breaks in the narrative and that close-ups did not represent heads severed from bodies. Other cultures, however, reportedly had great difficulties in following the syntax of living images (before World War II, that is, when our media companies began to colonize all perception worldwide). For this reason, it seems reasonable to assume that the ability to see image sequences followed from the historically acquired ability to follow not letter sequences as such, but rather letter sequences as imaginary image sequences. With lessons in silent reading, Europeans and North Americans would also potentially become subservient to cinema.

Schiller's ghost-seeing prince is obviously not enough to support such a bold thesis. Although he is a novel reader, we know nothing about his actual reading techniques, but rather only about his ideas of women, which sound as if they could only have come from Tasso or Ariosto. To conclude this ride through the history of literature and to provide a transition to chemically pure media technology, therefore, we will now consider Ernst Theodor Amadeus Hoffmann and romanticism in the narrowest sense of the word.

2.3.3.2 Hoffmann

Hoffmann's Gothic novel *The Devil's Elixirs* was published during the same period as the old monastery churches were being filled with ghost projections thanks to Étienne Gaspard Robertson and his silent mobile *lanterna magica*. The setting for the novel are the same old monastery churches, and it is about the same elixirs or drugs, namely a common visual hallucinosis, which will simply prove to be the correlate of a new silent reading technique. What remained only an empty promise with Schiller – ghosts would still appear to readers, while the author just decided to stop his novel – is taken literally or technically by Hoffmann. Not only do his fictional characters have repeated visions, but the entire novel also functions as an optical vision for all readers. The foreword by the so-called editor, which is de facto naturally that of the novelist Hoffmann, reads from beginning to end as follows:

Dearly would I take you, gentle reader, beneath those dark plane-trees where I first read the strange story of Brother Medardus. You would sit with me on the same stone bench half-hidden in fragrant bushes and bright flowers, and would gaze in deep yearning at the blue mountains whose mysterious forms tower up behind the sunlit valley which stretches out before us.

Then you would turn and see scarcely 20 paces behind us a Gothic building, its porch richly ornamented with statues. Through the dark branches of the plane-trees, paintings of the saints – the new frescoes in all their glory on the long wall – look straight at you with bright, living eyes. The sun glows on the mountain tops, the evening breeze rises, everywhere there is life and movement; strange voices whisper through the rustling trees and shrubs, swelling like the sound of chant and organ as they reach us from afar; solemn figures in broadly-folded robes walk silently through the embowered garden, their pious gaze fixed on the heavens: have the figures of the saints come to life and descended from their lofty cornices? The air throbs with the mystic thrill of the wonderful legends which the paintings portray, and willingly you believe that everything is really happening before your eyes. It is in such surroundings that you would read the story of Medardus, and you might come to consider the monk's strange visions to be more than just the caprice of an inflamed imagination.

Since, gentle reader, you have now seen the monks, their monastery, and paintings of the saints, I need hardly add that it is the glorious garden of the Capuchin monastery in B. to which I have brought you.

Once when I was staying at the monastery for a few days, the venerable prior showed me Brother Medardus' posthumous papers, which were preserved in the library as a curio. Only with difficulty did I overcome his objections to letting me see them; in fact, he considered that they should have been burned.

And so, gentle reader, it is not without fear that you may share the prior's opinion, that I place in your hands the book that has been fashioned from those papers. But if you decide to accompany Medardus through gloomy cloisters and cells, through the lurid episodes of his passage through the world, and to bear the horror, the fear, the madness, the ludicrous perversity of his life as if you were his faithful companion – then, maybe, you will derive some pleasure from those glimpses of a *camera obscura* which have been vouchsafed to you. It may even be that, as you look more closely, what seemed formless will become clear and precise; you will come to recognize the hidden seed which, born of a secret union, grows into a luxuriant plant and spreads forth in a thousand tendrils, until a single blossom, swelling to maturity, absorbs all the life-sap and kills the seed itself.

After I had with great diligence read through the papers of Medardus the Capuchin – which was extremely difficult because of his minute and barely legible monastic handwriting – I came to feel that what we call simply dream and imagination might represent the secret thread that runs through our lives and links its varied facets; and that the man who thinks that, because he has perceived this, he has acquired the power to break the thread and challenge that mysterious force which rules us, is to be given up as lost.

Perhaps your experience, gentle reader, will be the same as mine. For the profoundest of reasons I sincerely hope that it may be so. (Hoffmann, 1963, pp. 1–2)

I have cited this section in full to show how rigorously and systematically a romantic novel turns all questions about book technology into questions about image technology. Hoffmann, who neither only edited nor copied *The Devil's Elixirs* out of old books, but who rather first saw these stories, like all of his other tales, as colorful visions before his eyes (as described in his tale *The Sandman*) – this Hoffmann actually pretends to take the text from an old manuscript, whose unreadability refers back to the monastic practice of copying books prior to Gutenberg's invention. Hoffmann's readers, on the other hand, are not supposed to be bothered at all with the problem of decoding letters; the preface does not present them with that monkish manuscript, but rather with the landscape in which Hoffman read the manuscript for the first time. According to a lovely comment by Freud, "letters of the alphabet [. . .] do not occur in nature" (Freud, 1953–74, IV, p. 278); readers, whom Hoffmann can de facto naturally only lead under a dark plane-tree with his letters, therefore do not notice that they are reading. The training of silent and unconscious readers around 1800 won the first victory. The second and strategically crucial victory immediately follows: the landscape into which Hoffmann leads or seduces his readers is none other than the monastery landscape where the story itself will begin and also end. The readers who are transported to this landscape through the power of imagination therefore see the same church paintings and holy pictures that the protagonist himself also saw. And when these images "descend from their high ledge [. . .] to become alive," the readers are in exactly the same drugged or hallucinatory condition as the novel's protagonist: they also have "special visions" of a painted saint, who at the same time becomes the sole and incestuous love object of all of the novel's characters.

An optical ecstasy that Hoffmann only needs to acknowledge and award a good mark for: "Since, gentle reader, you have now seen the monks, their monastery, and paintings of the saints, I need hardly add that it is the glorious garden of the Capuchin monastery in B. to which I have brought you." In the context of these lectures, however, it is important to note the following: the fact that the narrative inventory of a romantic novel is exhaustively enumerated with "holy pictures, monasteries, and monks" emphasizes like nothing else that it is precisely these powers which allow this novel to become part of the image war. The prior of the monastery where the plot of the novel takes place and from which its autobiographical manuscript also comes wanted to "burn the papers." Hoffmann, on the other hand – only a decade after the great plundering of all monastic libraries in the *Reichsdeputationshauptschluss* (decision of the imperial deputation) on February 25, 1803, which transformed monastic knowledge into university knowledge and at the same time supplied the university library in Munich with its famous collection of manuscripts – ignores all prohibitions, which are obsolete because they are religious, and uses the self-same papers to create firstly his copyrighted novel and secondly his occupational sideline, which is protected by civil service law. The result is that the individual novel reader does not actually buy the "book formed from those papers" at all; rather, the reader acquires – and I quote – "the manifold images of the *camera obscura* that are revealed to you."

I can only say to that, like the mathematician, *quod erat demonstrandum*. It is simply written that German romanticism itself inherited the successful legacy of all Renaissance *camera obscuras* and all baroque *lanterna magicas*. This triumph came at a heavy price, though – namely, the fact that it halted media research in Germany for almost half a century because the historically awakened desire for images already appeared to be fulfilled in the imaginary world of the readers' souls – but I will discuss this later. When, to channel Novalis, "a visible inner world according to the words" of any author unfolds to the "right readers" who are trained in the new elementary school literacy techniques, or when, to channel Hegel, the collected history of western thought is reduced to a "gallery of images" after having run the gauntlet of Hegelian description, literature has arrived at the historic end of its monopoly on writing and it has caught up with all of the privileges of the *camera obscura* and the *lanterna magica*. Around 1820, the only remaining alternatives were either to perfect or technologize this magic. And technologize meant, as we will discuss in the next session, to remove the one fundamental deficit that the literary and thus imaginary *camera obscura* practically lived on. So that readers would continue buying romantic novels, it had to be absolutely impossible to store what Hoffmann called "inner faces" anywhere other than on paper, and they could therefore only be brought to life and pictured by the perfect literate reader and his inner world. The monopoly of writing was over from the moment that moving images could be transferred onto paper without any literary description or any help from a painter's hand (even if this hand was only tracing sketches made by a *camera obscura*). This break occurred with the invention of photography through Niépce and Daguerre, whose surname already contains the word for "war." *Post tenebras lux*, after darkness comes light, as Niépce's son named one of his polemics. But before we have a look at this light, which will incidentally bring forth its own entirely new types of darkness, infrared and ultraviolet, I would still like to make a few remarks concerning the film history of this romantic reading technique.

2.3.4 Romantic Poetry

It appears once more as if Virilio is the only one who recognized the relationship between romantic silent reading and film viewing. It would never have occurred to any of the Lutheran housefathers, who had to read suitable chapters out of the New Testament to their congregations every Sunday, to hallucinate the text like a projection from a *lanterna magica*. It was the silent and solitary reader who first carried out reading as a perspective on the visual information provided in the text (Virilio, 1989, pp. 36–8). In contrast to the hundreds of spectators in the peep show theater, however, this reader was absolutely alone. The reader's perspective could not be disputed by any fellow theater patrons occupying other positions in the auditorium, and for this and only this reason it could be completely believed, which means the illusion could be. As a French thinker, Virilio chooses the example of a newspaper reader in the Parisian metro to demonstrate that nobody likes it when others read over his shoulder. As a former Germanist, I should rather choose the classic example from Goethe's *Elective Affinities*, where this rule as well as its exception was formulated for the first time in 1809: Goethe's solitary reader Eduard made an exception to this rule, namely, when the person reading over his shoulder was Ottilie, who was at the same time object of a no less imaginary love.

It should not be too difficult to recognize, with Virilio, that the indisputable and imaginary perspective of solitary reading is a historical study of people's ability to perceive feature films and, to go a small step further, the exception to its rule of exclusivity is at the same time something like a preliminary historical study of the film star. The task that still lies before me is to eventually incorporate the pin-up girl into his concept, which requires a look past Hoffmann's foreword to the text of the novel itself. While reading alone in the monastery garden, the foreword states, holy pictures rise up from their ledges in the inner eye of the reader. It does not give too much away to say that the plot of Hoffmann's novel also talks about nothing else. The monk Medardus, around whose subjective perspective everything in *The Devil's Elixirs* revolves, in principle only falls in love with women who resemble a painting of Saint Rosalia installed in the monastery church. The historical basis of this confusion between heaven and earth is once more the Counter-Reformation and more specifically a painter who, as an ancestor of the novel's protagonist, took the baroque commission to create holy pictures that would arouse the sensuality of church visitors so literally that he chose the greatest whore in the world (also known as Venus) as the model for his Saint Rosalia. And even though the church expresses its thanks for this heinous deed by placing a curse on the painter's entire gender, it still does not prevent his holy-whore picture from continuing to be exhibited: apparently, even old European powers needed their pin-up girls.

Unfortunately, however, Medardus, the descendant of the painter and the novel's protagonist, does not reach the obvious conclusion that his entire lust for the flesh is an artefact of historical power. Rather, as a prototype of all romantic readers – and the women among you have surely already registered that Hoffmann's foreword is only addressed to men – the hero wants to make the holy picture of Rosalia congruent with a female co-reader standing behind his shoulder. It matters little that this beloved turns out to be his own blood relative. It is more important that Medardus, as the young woman's father confessor and religious instructor, pulls out all the stops to impose a readable underlying erotic meaning on the words of the Christian faith:

> [Aurelia's] presence, her nearness to me, even the touch of her dress set my heart aflame; the blood surged into the secret recesses of my mind, and I spoke of the holy mysteries of religion in vivid images whose ulterior meaning was the sensual craving of an ardent, insatiable love. The burning power of my words should pierce Aurelia's heart like shafts of lightning, and she would seek in vain to protect herself. Unbeknown to her the images which I had conjured up would grow in her mind, taking on a deeper meaning and filling her heart with intimations of unknown rapture, until at last, distracted with passionate yearning, she threw herself into my arms. (Hoffmann, 1963, p. 71)

In other words, while the novel works as a *camera obscura*, according to the foreword, the protagonist acts or seduces like a *lanterna magica*. The reproduction of images thus turns into image production. And this projection of erotic images into a woman's soul succeeds. Aurelia becomes the monk's bride. He is not actually able to seduce his bride in conversation simply because in the moment of embrace a chaste image of her double, Saint Rosalia, appears as a warning before his inner eye. But when Aurelia becomes a romantic solitary reader, there is almost nothing more standing between the plot of the novel and coitus, or between the worship of holy pictures and incest:

> For several days I did not see her [Aurelia], as she was staying with the Princess in a country residence not far distant. I could not stand her absence any longer and rushed to the place. Arriving late in the evening, I met a chambermaid in the garden who pointed out Aurelia's room to me. As I opened the door softly and went in, a breath of warm air perfumed with the wonderful scent of flowers dazed me,

and strange memories stirred in my mind: was this not Aurelia's room in the Baron's palace where I . . .

As soon as this thought struck me, a dark figure seemed to appear, and a cry of "Hermogenes!" went through my heart. In terror I rushed forwards, pushing open the door to the bed-chamber, which was ajar. Aurelia, her back towards me, was kneeling in front of a tabouret, on which lay an open book. I looked instinctively behind me. I saw nothing, and cried in a surge of ecstasy:

"Aurelia! O Aurelia!"

She turned round quickly, but before she could rise I was kneeling beside her, holding her in my arms.

"Leonard! My beloved!" she whispered.

An uncontrollable desire was seething within me. She lay powerless in my embrace; her hair hung in luxuriant tresses over my shoulders and her bosom heaved. She gave a gentle sigh. Savagely I clasped her to me; her eyes burned with a strange glow, and she returned my fierce kisses with even greater ardour. (Hoffmann, 1963, p. 203)

To understand the sexuality of our century one need only heighten this eroticism using media technology and replace readers with film-goers and film producers. In *Gravity's Rainbow*, Thomas Pynchon's great world war novel, there is a fictional expressionist director named Gerhard von Göll whose masterpiece is a UFA film with the obvious title *Alpdrücken* (Nightmare). At the high point of the film (in every sense of the word) Margarethe Erdmann, the star of all of Göll's productions, is tortured by a dark Grand Inquisitor of the Counter-Reformation (as you could already predict). Immediately afterwards, however, the so-called "jackal men" (disguised Babelsberg extras) come in to ravish and dismember the captive baroness. Von Göll let the cameras run right on. The footage got cut out for the release prints, of course, but (just like the transfer of the monk's manuscript from the monastery prior to the novelist E.T.A. Hoffmann) the original uncut version found its way into Goebbels' private film collection. Not only did the female star become pregnant at this literal high point of the film – and guessing the identity of the child's father became a popular party game (Pynchon, 1973, p. 461) – but also thousands of female filmgoers and even the girlfriends of filmgoers whose boyfriends were infected by the film. In conclusion, here is the experience of a V2 engineer upon seeing *Alpdrücken*:

He had come out of the UFA theater on the Friedrichstrasse that night with an erection, thinking like everybody else only about getting home, fucking somebody, fucking her into some submission . . . God,

Erdmann was beautiful. How many other men, shuffling out again into depression Berlin, carried the same image back from *Alpdrücken* to some drab fat excuse for a bride? How many shadow-children would be fathered by Erdmann that night? (Pynchon, 1973, p. 397)

According to the novel, the pin-up girls necessary for soldiers, which in many cases were and are stills from erotic films, delivered precisely these shadow-children to World War II and the *Wehrmacht*.

After this high point, we must leave literature along with all its illusions and shadow images to finally make a start on the archaeology of real images, that is, with the prehistory of photography. Before we do, I would only like to add that the holy-unholy pin-up girl in Hoffmann's *The Devil's Elixirs*, this double exposure of Rosalia and Aurelia, naturally did not yet have the mass media effect characteristic of our epoch. On the contrary: after all his failed attempts to seduce and sleep with Aurelia, Medardus becomes an absolute individual, namely a romantic author. He regrets his sinful worldly life, returns to the monastery that the temptation of Rosalia-Aurelia once lured him out of, and submits again to the authority of his monastic superior. As a practical act of repentance, however, he orders Medardus to do something that was entirely unheard of and impious in old European times – namely, simply to write the novel of his own autobiography:

> "Brother Medardus, I wish to impose on you what will doubtless seem like a new penance."
> I asked humbly what it was. He replied:
> "You are to write the story of your life. Do not omit a single incident, however trivial, that happened to you, particularly during your checkered career in the world. Your imagination will recapture all the gruesome, ludicrous, horrible, comical aspects of that life; it may even be that you will see Aurelia, not as Rosalia, the martyred nun, but as something else. Yet if the spirit of evil has really departed from you and if you have turned away from the temptations of the world, you will rise above these things, and no trace of them will remain."
> (Hoffmann, 1963, p. 319)

The prior's order brings the subject full circle for us. At the beginning of the novel, in the editor's foreword, readers were promised that they would see "the horror, the fear, the madness, the ludicrous perversity" of the life of Medardus before their own eyes like "glimpses of a *camera obscura*." At the end of the novel, the budding author and protagonist Medardus is promised exactly the same if he evokes his own life through writing. The religious authority thus paradoxically allows "the fantasy" Aurelia to be visually hallucinated again as a sexual object or pin-up girl, although in the meantime she has died a completely pious death. In other words, the church is no longer a church at all, but rather it has been transformed into romantic literature by internalizing all of the tricks of the *camera obscura* and the *lanterna magica*, and it functions as a visual hallucination for both its fictional author and its intended readers. The inner images, which this author sends, so to speak, as a *lanterna magica*, are received by readers as a *camera obscura*, because romantic literature as such is a magnetic for all eroticism.

Thus ends my short art history of optical media. I hope at least the literary section was entertaining because of all the ghosts and sexuality. But now please open or write a new, dry chapter, which will have to do without eroticism and women for a long time, simply because it concerns the European sciences.

— 3 —

OPTICAL MEDIA

3.1 Photography

3.1.1 Prehistory

From now on, there should be no more talk about *lanterna magicas* and *camera obscuras*, which only existed in novels like Hoffmann's as metaphors for poetic effects. To open a new chapter about the pre- and early history of photography, only the technical reality of these devices from the fifteenth and seventeenth centuries onwards is important. The *camera obscura* was one of the first technologies for receiving images, and the *lanterna magica* was one of the first technologies for sending images. The only thing that absolutely did not exist before the development of photography was a technology for storing images, which would allow images to be transmitted across space and time and then sent again to another point in space and time. For photography to emerge, it therefore needed an adequate channel (to return to Shannon's functions, although they still seem somewhat out of place here). Romantic literature was founded on the systematic exploitation of this channel's non-existence. If novels succeeded in giving rise to *lanterna magica* images in solitary readers, in principle these inner images still could not be stored – already proven by the success of novel sales. Since the time of Walgenstein, it was known that real images could only be stored when they – as in the case of plant leaves – were reduced to naked skeletons and then submerged in printer's ink. This demanding process of making nature print itself constituted the only exception to the rule that the storage of images had to go through the two intermediate stages of the human eye and the human hand and thus become painting and art. And as was shown in the section on theory, the interface called

118

the human eye always introduces the imaginary into images because of its ability to pick out shapes in a world view infiltrated by accident and noise. Instead of storing the empirical probability distributions of lights and shadows, modern painting, like modern literature, presented its public with the idea of a subject, and thus of an artist. This is what Heidegger called "the age of the world picture." And if this idea superimposed the image of a saint and that of a whore on one and the same painting, the imaginary was perfect.

No perfecting of painting would therefore have been able to make the transition from visual arts to optical media. In spite of all beliefs in progress, there is no linear or continuous development in the history of media. The history of technologies is, on the other hand, a history of steps or, as stated in Thomas Pynchon's novel *V.*, "History is a step-function" (Pynchon, 1990, p. 331). For this reason, Goethe's great fear, which he revealed in a 1797 manuscript with the remarkable title *Kunst und Handwerk* (Art and Craft), could not happen at all historically: namely, that painting would simply be overrun by machines, that painting techniques would be mechanized, and that countless identical reproductions would replace the unique original. Machines are not just simple copies of human abilities.

3.1.2 Implementation

In the case of photography, the historical step amounts rather to a painting mistake or offence that became the foundation of a new scientific media technology through the re-evaluation of all values, as Nietzsche would have said. Do not confuse this literal perversion with Hegel's dialectical negation, where a higher philosophical truth emerged from a double negative and the book of books, Hegel's own philosophy, emerged from the abolition of all other books. The re-evaluation of all values simply means transposing a sign so that a negative becomes a positive or, to formulate it in images of photography itself, a positive becomes a negative.

The negative of all painting existed in its naked materiality, namely in its colors. It therefore existed neither in symbolic meaning nor in the imaginary effects of red, green, or blue, but rather in the simple reality of pigments, as they have been known since time immemorial. I recall carmine red, Prussian blue, lapis lazuli, etc. I recall above all the last great European novelist, who conceived of himself as a magician or an "illusionist." Humbert Humbert, the protagonist of Nabokov's most widely read novel, talks about Lolita, himself, and art in the very last sentence: "I am thinking of aurochs and angels,

119

the secret of durable pigments, prophetic sonnets, the refuge of art. And this is the only immortality you and I may share, my Lolita." (Nabokov, 1958, p. 311)

In other words, pigments only last in art. But in this miserable world, which only becomes even more miserable through reproduction or copying, not only do the Aurelias and Lolitas become older, but so do their images. Because he conceived of this, Nabokov rises above his romantic predecessors.

When and as long as pigments shone from paintings, as artists had applied them to canvas using oil-based paints since 1450, or at the very latest since the time of the Van Eyck brothers, everything was aesthetically in order. But paintings do not always hang in museums where light, air, and temperature are technically filtered and optimized; in unfavorable locations they fall victim to the sun or the darkness. Then the aesthetic negative comes into play: many colors become brighter or darker than when they were first applied, and many of them turn into other colors (like the American color television standard). Painters knew this from bitter experience, because it made artworks intended for immortality suddenly mortal. Since the Renaissance, therefore, warnings against bad pigments or chemical colors, like dragon's blood, lac, vermilion, and carmine, which darken or whiten afterwards (Eder, 1978, p. 85), have stood alongside recommendations for better colors in painting instructions. But it never occurred to any of the painters who had discovered perspective and the *camera obscura* to turn this handicap into an asset by taking advantage of the whitening or darkening effect itself. As far as I know, there is not a single fictional painter in a novel or tale who took revenge on an evil client by intentionally using deteriorating pigments so that the image depicted on the painting literally faded away after 30 years. One can therefore venture the thesis that there has been no way to go from aesthetic experience to media technology in the past, nor can there be in the future. This does not make the reverse untrue, however. Rather, there is at the same time a second valid thesis that there are undoubtedly ways to go from aesthetic handicaps to media technology, even ideal ways. Just as technical media like the telephone and the gramophone were invented in the nineteenth century for and by the deaf, and technical media like the typewriter were invented for and by the blind, so began the first experiments with the darkening or lightening of certain chemicals in the seventeenth century, which directly led to photographic film through the work of Niépce and Daguerre. Cripples and handicaps lie like corpses along the technical path to the present.

What follows now are some of the highlights of the "deeds and sorrows of light," as Goethe called optics in his *Theory of Colors*. It was already known to the advanced civilizations of antiquity that light can bleach colors as well as canvas (the painterly equivalent of photographic film). But seventeenth-century natural sciences first made it clear that the normally green color of plant leaves, at least before Walgenstein submerged them in printer's ink, is no accident and also does not stem from warmth but rather is produced exclusively through the influence of light. Chlorophyll was thus the first storage device for light ever discovered, yet it worked from nature and was therefore not manipulable (Eder, 1978, p. 55).

Scientists were not the only people who conducted research on light. Just as magic and secret societies emerged during the twilight of the *lanterna magica* and its religious propaganda, so were many photochemical discoveries byproducts of alchemy experiments that were neither planned nor desired. As you know, absolutism was not based on paper or computer money like today's high-tech empires, but rather it had its gold currency and its newly invented or permitted national debts, which were caused by its constant need for precious metals. In practice, alchemy accordingly wanted to make gold or silver out of cheap materials like kaolin, which by an accidental discovery then led to porcelain. For example, a civil servant from Saxony who wanted to "capture the world spirit" long before Hegel, that is, using alchemy rather than philosophy, accidentally discovered a chemical substance that could store light and transmit it again in darkness. An unsuccessful salesman continued the experiment and discovered, with even more luck, the chemical element phosphorus or "carrier of light" (Eder, 1978, p. 58). With phosphorescing substances magicians, secret societies, and con-artists could then make their ghosts or skulls glow.

Another discovery was crucial for the development of photography: in 1727, Dr. J.H. Schulze, a professor of Greek and Arab languages in Halle, took up the experiment with phosphorus again, but in the good alchemical tradition he also introduced silver into the experiment. When he accidentally performed his tests in front of a sunlit window, Schulze discovered that the silver salt lying in the sunshine became dark, while the silver salt lying in the shade remained bright. And just as Kircher planned to transmit actual secret characters over the head of an unsuspecting enemy with the *lanterna magica*, so did Schulze use the light sensitivity of silver salts to code data. He wrote dark letters on a piece of glass, placed it between the sun and the silver salt, and in this way achieved the first photographic negative: all the brightness disappeared from the illuminated chemicals everywhere where the light was not filtered out by the letters. Unfortunately, Schulze's experiment also proves at the same time that the idea of photography was still impossible in the eighteenth century: Schulze did not want to store the contingent nature of the real (in Lacan's sense of the word) in a technical medium, but rather he wanted to introduce the symbolic, namely a written code, into nature. When historians thus claim that Schulze's light writing already anticipated photography (Eder, 1978, p. 62), this is actually valid for the word "photography" but not for the process itself, which is the object of all technical media.

Due to time constraints, I will not pursue the history of photography in all its detail, even if such a thing were possible. For our purposes, it must only be fundamentally clear that the discovery, use, and optimization of light sensitivities were linked to the general history of the origins of chemistry in the eighteenth century. This research was impossible so long as the four Greek elements of fire, air, water, and earth were considered the only components of all being. To be able to isolate a photochemical effect as such, chemically precise distinctions like those between fire and light or between light and heat first had to be made. For example, Beccaria, the great legal reformer who also experimented with silver salts, had to learn with great difficulty that it is not warmth but rather light that blackens these salts. You can read about other names and chemical discoveries, which I will skip over, in the old-fashioned but positivistically exact history of photography produced by the imperial and royal counsellor Dr. Josef Maria Eder in 1905.

With the development of optical lens systems, as I have previously mentioned, Huygens arrived at the fairly adequate thesis that light consists of waves, which naturally were not defined as electromagnetic waves prior to the work of Maxwell and Hertz, but rather as elastic waves of an undefined medium, as the great mathematician Euler called them. However, after 1700 – since the formation of a Royal Society in London with Sir Isaac Newton as its president – the opposing thesis gained acceptance and light was defined as a mass of tiny bodies or a particle stream. Newton had reconstructed the classical *camera obscura* experiment with one refinement: he placed a glass prism in the beam of light between the sun, the hole in the chamber wall, and the dark projection wall. Since Descartes had already made the natural rainbow (if not also gravity's rainbow) calculable (see the historical part of Goethe's *Theory of Colors*), the result of Newton's experiment was the first artificial rainbow. The simple white sunlight dissolves into countless colors of the spectrum between violet and red, from which Newton inferred that light consists of parts or particles and that a convex lens would combine many colors to make white again. The experiment with the lens succeeded, and for the first time it compared scientific color synthesis with painterly color synthesis. When mixing oil-based paints the sum always becomes darker because light disappears and consequently a subtractive color synthesis takes place; since Newton, however, there is also additive color synthesis where the sum of many colors is brighter than its addends. We will have to come back to this when we discuss color television.

At the turn of the century around 1800, after chemistry and physics were also established as academic sciences, the chemistry and physics of light – Schulze's photochemical effect and Newton's spectrum analysis – intersected for the first time. No less a person than Friedrich Wilhelm Herschel, the son of a Hannover court musician who had risen to become a British mathematician and court astronomer, introduced the new distinction between light and heat to photochemistry as well: he proved that for the human eye the broken sunlight in Newton's prism actually stops at red and then turns to black, but on a storer of heat like the thermometer it also has measurable effects beyond red. In other words, Herschel discovered infrared as a physical embodiment of heat and thus also as a medium on which night vision aids for *Waffen-SS* tanks in World War II were based, and which is still the basis for tactical anti-aircraft rocket sensors today.

After Herschel's experiment became known, Johann Wilhelm Ritter, the physicist among the German romantics, took a mirror-image step. Ritter posed the very methodical question: if the solar spectrum is brightest in the middle but warmest at its end, then why should cold light not also exist beyond the other end of the visible spectrum? To answer his question, Ritter clearly needed, as he wrote in his 1801 book *Bemerkungen zu Herschels neueren Untersuchungen über das Licht* (Remarks on Herschel's Recent Experiments on Light), a chemical "reagent that has its strongest effect beyond the violet in the same way that our thermometer has its strongest effect beyond the red" (Ritter, 1986, p. 119). Thus it was that Newton's solar spectrum and the light sensitivity discovered by Schulze and further researched by the chemist Scheele came together in a single transparent experiment. Ritter was able to show that when a beam of sunlight is split using a prism, the silver chloride is blackened the most and the deepest by those parts of the spectrum that our eyes can no longer perceive as light. Ritter thus discovered ultraviolet light, which provides light without heat – an analog to Herschel's infrared light, which provides heat without light.

As the everyday concept of light expanded symmetrically towards its two invisible edges, however, Newton's particle theory also collapsed again. The wave theory posited by Huygens, and still only supported by Euler and Kant in the eighteenth century, was revived in 1802 by Thomas Young, who added the new point that extremely long light waves are infrared and extremely short waves are ultraviolet. This concept of waves or frequencies, which Euler had also introduced in modern music theory, in contrast to Pythagoras, allowed researchers like Fresnel or Faraday to study light interference and its moiré-type pattern around 1830, which would be important for fundamental film effects.

It is important to explain that such theories were scandalous in an epoch when Germany's prince among poets was working on his theory of colors in total opposition to Newton and all of the natural sciences based on his work. For Goethe, light was an urphenomenon of nature, and Mother Nature could not be reduced to individual parts or waves – this would amount to sadistic incest – but rather always had to be described or worshipped only phenomenally. According to his theory of colors, each of the different colors emerged as a more or less equal mixture of both primary conditions, light and darkness. In other words, colors were dialectical effects of the polarity between God and Mephisto, Goethe the poet and Goethe the civil servant. The prince among poets would never have tolerated the notion that light – like Ritter's ultraviolet – could be at its maximum where natural human organs – like Goethe's beloved eyes – in principle no longer suffice. The romantic physicist Ritter was never able to recognize this death in his discovery. Ritter's essay about ultraviolet concluded with the dramatic formula that it would soon be possible to trace all of the polarities of nature – electricity, magnetism, and heat – back to a single identical principle and to find this principle embodied in light, for "light is the source of any strength that creates life and activity; [light] is the seed that produces everything good on Earth" (Ritter, 1986, p. 127).

In deed and in truth, Ritter's ultraviolet led to the discovery of X-rays barely a century later, which then showed the protagonist of Thomas Mann's *The Magic Mountain* what it means to be already able to see one's own death while one is still alive, namely in X-ray photographs of a tubercular lung. Worse still than Thomas Mann's media consumer panic, however, were an irony of media history and an effect of theory. The irony consisted in the fact that not only did the fictional Hans Castorp die of tuberculosis, but also the poor romantic physicist Ritter, who discovered ultraviolet. The effect of the theory of invisible light, on the other hand, was the total elimination of the ancient visibility postulate. When the great Viennese physicist Ludwig Boltzmann (whose entropy formula is mathematically identical to Shannon's later information formula) wanted to prove that there is no thought without a corresponding material, namely brain-physiological apparatus, he suggested for future research that our thoughts should simply be projected onto an X-ray screen (Boltzmann, 1979).

Just as Boltzmann's opponents in physics, like Ernst Mach, still had the upper hand during his lifetime and his own atomism only triumphed posthumously after he shot himself, Goethe's theory of colors also dominated over German optics during his lifetime. Romanticism meant – according to the thesis of these lectures – transferring all of the optical real into the imaginary world of the readers' souls, where it naturally could not be stored.

3.1.2.1 Niépce and Daguerre

This lecture will therefore cut to another new scene – this time imperial or Napoleonic – as we step onto French soil. This requires, though, that – quite in the manner of Napoleon – we remain on German soil for a moment longer in order to carry off art treasures and technologies and to destroy the Holy Roman Empire.

As we know, early modern empires were based on printing, which in turn was based on paper for books and older parchment for imperial or royal documents. At the end of the eighteenth century, however, not only did the state experience a great revolution, but paper also experienced a revolution. Following the model of animal hide parchment, which was in principle finite, paper had for centuries been created in finite large sheets, which then had to be folded, cut, printed on, and bound together in books. Folios, quartos, and similar old book names all originated from such discrete formats. From 1799 onwards, on the other hand, there existed paper machines that produced endlessly long ribbons, like idealized toilet paper, and since the invention of the high-speed printing press in 1811 there was also a printing technique that corresponded to this new format or unformat: rotary printing. Gutenberg flat relief plates no longer printed letters on an equally flat and limited paper sheet, but rather an endlessly turning print cylinder revolved over an endlessly long paper cylinder, which is unrolled underneath it. To the film scholar in us, this technical innovation naturally evokes the image of rolls of film, and it reminds us how Orson Welles immortalized the rotary press in his film about a not very fictional newspaper mogul named Charles Foster Kane. In the middle of the nineteenth century, however, the rotary press first ensured that old European empires gradually became democratic states with unlimited paper supplies and newspapers that published an unlimited variety of opinions. Of course, the first rotary presses, in London in 1811, stood in newspaper printing works, and they started modern mass journalism. Heinrich Bosse called it "writing in the age of its technical reproducibility" (Bosse, 1981), a type of writing incidentally, that according to Bosse's proof, also led to the new legal construction of an endless copyright.

But now what had already been demonstrated for the Gutenberg era repeated itself: each historically defined print medium needs a corresponding optical medium. Gutenberg's printed books called for woodcuts and copperplate engravings. The rotary press needed something that was called an illustrated newspaper in the middle of the nineteenth century; today it is simply called an illustrated. To answer this need, before it was finally met by photography, a certain Aloys Senefelder invented lithography in 1796.

With respect to the technical principle of lithography, which first made newspaper illustrations possible in mass editions, it suffices to say here that in contrast to gravure printing with copperplate engraving and relief printing with woodcuts it is planographic: as with photography there is only a single plane, which is soaked partly with fat and partly with water and then pressed. Because water and fat normally do not mix, Senefelder was already able to distinguish dark and bright places in the image very well. In 1827, his successors even achieved what eighteenth-century copperplate engravers were only able to realize with tremendous effort: hassle-free four-color printing, which enabled reproductions of artworks – or kitschy oil paintings, to be more precise – to be brought for the first time into every middle-class living room. Walter Benjamin's work of art in the age of its technical reproducibility found its material basis in lithography.

Theoreticians were not the only ones to profit from Senefelder, but also Napoleonic counts. In Munich in 1812, Count Philibert de Lasteyrie-Dusaillant, a son-in-law of the famous General Lafayette, studied lithography under the inventor himself. The unfortunate Russian campaign drove him to France, and it was from there that he first succeeded in importing a lithographic machine to Paris under the Bourbon Restoration (just as King Ludwig XI once brought the Gutenberg printing press to his emerging nation-state).

126

And now we have really managed to enter the realm of media studies itself, for the technique of photography was developed around this lithographic machine through the teamwork of two men: Niépce and Daguerre.

Herewith two brief introductions: Joseph Niépce was born near Chalon-sur-Saône. He was originally supposed to become a priest, but as a result of the French Revolution he pursued a carrier as an officer instead. The victors of the Counter-Reformation became combatants in the national wars. On a whim, Niépce accordingly adopted the very unchristian name Nicéphore or "victory bearer" (Jay, 1981, p. 11). One of Niépce's brothers also built an internal combustion engine to power a boat – a motor, incidentally, that despite the patents granted by Napoleon was suspected of being yet another dream for the old phantasm of a *perpetuum mobile*. You can see from this that the Niépce brothers were already chasing desperately after the dream that Edison first realized: the invention of invention itself. If you would like to know more, go to Chalon-sur-Saône. In the center of the city you will find, in addition to good wine, a museum of photography dedicated to Niépce. There you will also see the construction plan of a *pyréolophore* or internal combustion engine, which as a precursor to all of our cars and tanks was supposed to have made it possible for the first time ever to put a submarine in the Saône.

Louis Jacques Mandé Daguerre, whose war-like name I have already emphasized, was not related to generals, like Niépce, but rather to civil servants. Daguerre himself, on the other hand, began as a painter, and he displayed a particular talent for perspective and lighting. For the sake of the imaginary in painting he changed over to that hybrid mixture of arts and media that was introduced in the nineteenth century. Daguerre first painted so-called panoramas, giant paintings that surrounded their spectators with perspective on all sides of the horizon. In 1822, he personally developed the diorama in Paris. As the name already suggests, this was a panorama that was partly reflective and partly transparent and thus combined traditional painting with a *lanterna magica* effect. So long as Daguerre's diorama was presented to a paying audience with reflected light, for example, it showed a peaceful daytime view of Vesuvius, but when light shone through the image, then a night-time view of the same volcano appeared suddenly from the back of the canvas – with eruptions, fire, lightning, and multicolored illuminated clouds. Daguerre thus modernized painting according to that apparent movement that Alberti began with his perspectival day and night views of nature, and that Schröpfer or Robertson had further developed with the *lanterna magica*. And finally, because the battle between Enlightenment and superstition was still raging during Daguerre's lifetime, the minor miracle occurred that a church of all things bought one of Daguerre's dioramas for its own illusionistic purposes (Eder, 1978, p. 211). Daguerre was therefore predestined from the start, much earlier than Niépce, to implement the phenomenon called photographic world view.

After these brief introductions, we now move on to the collaborative teamwork of these two founders. In the beginning, it was Niépce alone who (with more success than his brother, the *perpetuum mobile* constructor) was simply looking for a method of perpetuating images of nature – with the express goal of automating the production of lithographs, whose existence he had first heard of in 1813. Niépce's heliography, or the art of sun writing, was supposed to serve the same function for the multimedia system of rotary press and lithography as Walgenstein's self-printing of nature served during the Gutenberg era. He furnished the sun with the ability to etch images of illuminated objects, which were themselves typically already images, onto a metal plate without the operation or interposition of a painter's hand. These metal plates could then profitably replace Senefelder's slates. In a series of experiments, which really tested all of the light-sensitive materials known at that time, Niépce discovered that the most suitable chemical was asphalt, a substance that in the meantime has successfully covered half of the nation's ground. But even with asphalt, it took hours or days until the copperplate of a Renaissance cardinal was recorded and its black-and-white values were also developed and fixed in Niépce's *camera obscura*. It was therefore purely a technology for reproducing images, and new recordings of so-called nature, of chance itself, were practically out of the question simply because sunlight and shadows do not always stand as still as they once did in the Old Testament (Eder, 1978, p. 223).

In place of precisely this deficit came Daguerre – through the intermediary of a Parisian *camera obscura* dealer, by the way. As a painter and illusion artist, what was important to him was not primarily eternal storage and reproduction, but rather, as with his diorama, the recording of fleeting events or time processes. Through a formal legal contract, Niépce transferred all of his secret technical knowledge to Daguerre, who for his part brought nothing further to the contract than the joy of experimentation and patience, with which he survived Niépce's death in 1833 and was celebrated as the sole inventor of the typically named daguerreotype.

128

Indeed, daguerreotypy in 1839 possessed only the slightest similarities to Niépce's original project. According to expert advice, it
was an illusion to hope for any model for lithography from it,
simply because the delicate photographs would be "destroyed past
redemption" if "subjected to the pressure of a roller" (Eder, 1978,
p. 237). On the other hand, there were two immense advantages that
compensated for this lack of robustness and reproducibility. First,
Daguerre no longer employed symbolically coded models like copperplate engravings, in which all of the real or the noise then fell on the
part of imperfect heliography. Instead, he began with real conditions,
which naturally changed with the sunlight, and he therefore had to
make substantial improvements to the recording speed: in "midday
climate," the great physicist and expert Arago estimated that in the
climate of midday two to three minutes would be sufficient to make
daguerreotypes of nature. The fundamental trend of modern media
technologies to replace static values with dynamic values and to
replace steadfastness with speed had, after Claude Chappe's optical
telegraphy, now also caught up with optical image storage. According to Daguerre's contemporaries, his process was "60 to 80 times"
faster than Niépce's (Eder, 1978, p. 228).

Even if it appears quite old by today's standards, which are measured in micro- and nanoseconds, Daguerre reportedly accomplished
this sensation through two accidental discoveries. The first accident
occurred one day during an experiment when a silver spoon lay on an
iodized silver plate. When the sun shone on it, an image of the spoon
appeared on the plate, whereupon Daguerre naturally said farewell
to Niépce's asphalt. The second accident occurred when:

[a] number of plates he had previously experimented upon in the
camera obscura had been put aside in an old cupboard and had
remained there for weeks without being further noticed. But one day,
on removing one of the plates, Daguerre to his intense astonishment
found on it an image of the most complete distinctness, the smallest
details being depicted with perfect fidelity. He had no idea how the
picture had come, but he felt sure there must be something in the
cupboard which had produced it. The cupboard contained all sorts of
things: tools and apparatus, chemical reagents, and among the other
things a basin filled with metallic mercury. Daguerre now removed
one thing after the other from the cupboard, with the exception of
the mercury, and still he regularly obtained pictures if the plates which
had previously been submitted to the action of images in the *camera
obscura* were allowed to remain for several hours in the cupboard. For
a long time the mercury escaped his notice, and it almost appeared

to him as if the old cupboard were bewitched. But at last it occurred to him that it must be the mercury to whose action the pictures were due. A drawing made with a pointed piece of wood on a clean pane of glass, remains invisible even to the most acute sight, but comes to light at once when breathed upon. The condensation of the water vapor (deposited in small drops) differs in the parts touched with the wooden point and those left untouched, in the same manner as took place in Daguerre's pictures. (Eder, 1978, p. 228)

As a research method, this mixture of accidents and the systematic elimination of accidents requires a brief comment. First, I hope it is clear how much chemistry must have historically already taken place in order that iodized silver and quicksilver could be accidentally placed in the same cupboard. After all, even the surreal accident that Lautréamont later defined as the encounter of a sewing machine and an umbrella on a dissecting table presumed the existence of three technical inventions. Once they exist, however, artificial substances or machines are able to react to one another without human intervention, just as today's random processes between silicon and silicon dioxide, which computers ultimately consist of, relieve people of all thinking. Second, like Ritter's discovery of ultraviolet light, the history of Daguerre's invention very clearly shows how methodically theoretically preset values were investigated after 1800. That nature could be brought to create a black-and-white image of itself was a result of sheer will and not an assumption that was already justified elsewhere. Only the preliminary decision to already presuppose this possibility ensured that even apparently bewitched cupboards no longer raised any suspicion of witchcraft or magic at all. In other words, if an accidental effect like the one that occurred in Daguerre's cupboard had taken place 200 years earlier, which is not entirely out of the question, the whole matter would have fallen flat again simply because no one would have captured, stored, recorded, and exploited it as a natural technology. Niépce and Daguerre thus represent the beginning of an epoch where the duration, the reproducibility, and practically even the success of inventions – and in the end that means historically contingent events – are guaranteed. In 1839, we are still not quite in the epoch where President Reagan could solemnly announce as the emperor of California and in the name of holy Hollywood that in the future American inventors can or must invent a surefire protective barrier to defend against all missiles that want to attack California from the evil empire; but we are also no longer far from this epoch.

This new guarantee of inventability in general also corresponds to the public reception of Daguerre's invention. Up until now – namely, up until the solitary death of Niépce, the founding hero – these lectures have had to report on more tragedies and testimonials to poverty, but we are now coming to brighter times and the worship of bourgeois geniuses, even if they are only junior partners in inventor teams, like Daguerre. The most famous physicists and academics in France, above all Arago and the gas researcher Gay-Lussac, supported the inventor and earned him the Cross of the Legion of Honor and an annual life-long pension of 6,000 francs as a "national reward" (as it is called in the specially enacted law) from the "Citizen King" Louis Philippe (Eder, 1978, pp. 232–3). Daguerre therefore died as a successful man who not only earned 400 gold francs for each of his cameras and demanded 60 to 120 gold francs for his daguerreotypes (Eder, 1978, p. 253), but who could also make money by relinquishing his patent rights. This enabled Daguerre to afford a park for his retirement estate, which contained small lakes, waterfalls, and above all (as if to prove the thesis of these lectures) the ruins of a castle and a gothic chapel (Zglinicki, 1979, p. 149). For France, on the other hand, all of Daguerre's honors took place with explicit reference to modern copyright law, which the French Revolution (under the overall control of Lakanal, the telegraphy advocate) had invented for "spiritual heroes" and especially for novelists in 1794. By panning from literature to media technology, his invention proved in retrospect to be the invention of invention itself.

The historical goals that were initially prescribed for daguerreotypy can be gathered from the ceremonial speeches about Daguerre that were held in his presence. On the one hand, Arago emphasized that during Napoleon's Egyptian campaign art history and archaeology already urgently needed daguerreotypes to reproduce artworks and inscriptions without the falsifying hand of the painter. (Even the ten-volume description of Egypt that was edited on Napoleon's behalf by none other than Baron Joseph Fourier, the founder of modern wave theory, was still not a technical medium.) On the other hand, painters could also employ daguerreotypes as cheaper and at the same time more precise replacements for their usual sketches and models. This was supposed to have far-reaching consequences for the invention of film, but it also already happened, for example, when the famous Ingres made his famous painting La Source (The Spring). The painting shows a stylized Greek woman carrying a pitcher; however, the photograph Ingres used as a model shows – or rather conceals – an ordinary Parisian prostitute, as they were typically used as models at that time. So much for the new competitive relation-ship between art and media technology, painting and daguerreotype, which Arago only touched upon. To the scientist, as expected, the "scientific advantages" of invention were far more important (Eder, 1978, p. 237).

In this regard, the first and most dramatic point that Arago emphasized was the possibility of absolutely measuring light and its strength. With human eyes alone there were always only rela-tive estimates of the strength of light, but the daguerreotype did for photometry precisely what the Parisian standard meter did for land surveys and cartography in 1790. Lambert's photometry thus freed itself from all phenomenological or subjective appearance in order to become a technically precise form of measurement. A second point was the sensitivity of the new medium, which made light more intense compared to the eye and which made it possible to force the moon or the rings of Saturn to produce experimental effects. And finally, because their exactness surpassed all other arts, Arago also men-tioned the possibility of measuring daguerreotypes themselves with rulers and compasses to infer the length and angle ratios of the objects represented. From this emerged Oliver Wendell Holmes Sr.'s great idea to destroy all artworks (for instance, using gunpowder) immedi-ately after they are photographed, and since the two world wars the even greater idea has emerged of airplane or satellite reconnaissance photographs that finally contribute to the systematic destruction of the photographed object . . .

3.1.2.2 Talbot

According to the honest Arago, there are only two things in the world that balk at daguerreotypes: first, paper, on which all writing and books are based, and second, people, on which all inventions were based according to early capitalist doctrine. Daguerreotypes could therefore only be printed through the mediation of a lithographer or copperplate engraver, and people, who are much too restless, could only be captured with extreme difficulty. In 1839, in the land of unlimited possibilities, Professor Draper and Professor Morse simply made a person sit for half an hour in the blazing sun with white face powder and closed eyes until the first portrait photograph was taken or rather waited for (Eder, 1978, p. 271). Their studio in New York was thus the first photographic portrait studio in the world and the first technology for truly storing human faces, and I cannot avoid making special reference to Morse's involvement: Samuel Morse was

132

the painter who in 1837 had made the electric telegraph – the all-important modern broadcast technology – ready to go into production. There are no media, but rather always only multimedia systems.

As you can imagine, however, the multimedia system between photochemical storage, on the one hand, and paper and people, on the other hand, wanted to be completed not only in the American sense, like Samuel Morse's telegraph standardizer, but also formally. Gradual changes to Daguerre's process, all of which I cannot deal with here, eventually led to instantaneous photography, which made it possible for the first time to contemplate the storage of moving images, which means film. I will only name the most important of these steps and researchers.

William Henry Fox Talbot, Daguerre's previously underestimated English competitor, produced the first photographs on paper. During his trip to Italy, this rich Englishman repeatedly attempted to capture beautiful romantic images with the *camera obscura* and then sketch them by hand, as was typical since the time of the Renaissance. But the successful perspectivization of nature, which had filled all of the painters since Dürer with pride and contentment, was no longer satisfying after Daguerre made the physically real storable. Talbot, the failed painter, thus sought and found a technical method of photographing directly onto paper soaked in iodized silver and silver nitrate. From the *camera obscura* he developed the camera in the modern sense of the word. The hands of lithographers, which had previously stood between technical recording and book reproduction, were thus no longer necessary. In the entire operation of new optics there were no longer any arts, but rather only media. Talbot himself, who called his calotypy or "beautiful impression" technology "the pencil of nature" as it allows nature to inscribe images of itself, actually used it – strictly according to McLuhan's law that the content of media are always other media – for magnificent volumes of beautiful images of European nature or art, but the path to illustrated magazines based on photographs and to all the magazines sold under the counter was already mapped out.

Another of Talbot's innovations, the introduction of negatives, was relevant in an entirely different way. Unfortunately, I do not know if the concept of positive and negative in photography was derived from the much older analog concept of mathematics or the concept of positive and negative in the theory of electricity, which was only 50 years old at that time, but this should be clarified at some point. With Daguerre's process, the recording and fixing of the image always produced a positive reproduction of the lighting conditions.

133

Experiments with negative reproductions failed because of their irreversibility. Talbot resolved this problem by always photographing the negative again just as it had been produced itself in order to obtain as many positive copies as he wanted. This had two far-reaching effects, one hidden and one apparent. The secret effect was that photography was not only capable of amplifying or magnifying the light, for which Arago had already praised Daguerre, but rather since Talbot it was also capable of magnifying or reducing the size of the image when copying the negative. And because the whole world always only believes in the value of magnification – I am reminded here of Antonioni's *Blow Up* – it should be emphasized that image reduction also has strategic effects. For example, in 1870, several weeks into the German-French War, Paris was surrounded on all sides by Moltke's telegraphically controlled armies. The urgently needed exchange of messages with the republican armies, which had repeatedly tried to relieve the sacred heart of France, seemed impossible until they had the idea of using a photograph. They photographed the letters that were supposed to go from Paris to the armies, reduced the size of the copy, photographed the copy, etc., until the entire text fitted on the foot of an innocent carrier pigeon, which flew to Gambetta's general staff. If German falcons did not intercept the pigeon, the unreadable text was decoded "with the help of an electrically illuminated magic lantern," which projected a magnification of the photograph on a readable screen (Ranke, 1982, p. 49). Twentieth-century intelligence agencies developed techniques of reducing entire secret messages to the size of a harmless typewriter point.

The obvious effect of Talbot's innovation requires less commentary. The consequences of unlimited copying are clear: in a series first of originals, second of negatives, and third of negatives of a negative, photography became a mass medium. For Hegel, the negation of a negation was supposed to be anything but a return to the first position, but mass media are based on precisely this oscillation, as it logically calculated Boolean circuit algebra and made possible nothing less than the computer.

Only Talbot, who had done for optical reproducibility exactly what Gutenberg had done for printing technology, was a British snob who hated copying. Like Gutenberg, who had demanded (because of a Strasbourg trial in which he had become entangled) that a hole should be bored into his printing press, that a large stake should be stuck in the hole, and that the unmovable and unusable machine should then be laid in a grave, Talbot also prosecuted everyone who wanted to copy his method. Only a threatening intervention by the

Royal Society was able to induce him "to take a less severely obstruc-
tive attitude in the interest of the arts and sciences" and renounce his
patent rights with the lovely exception that commercial uses would
continue to remain forbidden (Eder, 1978, p. 321).

For the sake of simplicity, let us assume that it was this immediate brake on exploitation that brought so many other photographic
materials into the world between Talbot's paper and the photographic final solution (namely, Eastman's celluloid), such as those
terribly heavy and fragile photographic glass plates that a dragoon
lieutenant and a cousin of Niépce's developed in 1847. Glass actually plays a fundamental role in photography, but as a lens system
rather than as a storage surface. The reason for the low light level
of the lenses employed by Daguerre and Talbot was that people had
to sit still for half an hour in the blazing sun until their portrait was
in the box. To rectify this shortcoming, an experimental search for
faster lenses began, which became a true science in Abbe's Zeiss
works in Jena. A search for light-sensitive emulsions also began,
which did not end until the development of digital photography. To
bring about the conditions necessary to make feature films possible,
the new optical medium eventually had to separate itself again from
the traditional printing press, into which it had been so effectively
integrated thanks to Talbot. In other words, it had to obtain its own
materiality, something besides metal, paper, or glass – namely, that
strange half-transparency we call film without even hearing it as a
foreign word at all. I have looked it up, and I am happy to report
that the Anglo-Saxon word *aegfelma* or egg skin and the Old Frisian
word *filmene* or soft skin are joined to the West German *felmon*,
meaning skin. The root of all of these words is naturally *fell* or fur,
but since 1891 such animal furs of the parchment variety are gone;
the advancement that replaced them was light-sensitive film for photographs and sequences of images. However, Kluge's *Etymologisches
Wörterbuch der deutschen Sprache* (Etymological Dictionary of the
German Language) unfortunately draws no parallels between film
and parchment.

The other happy news concerns the inventor of film, who brought
the same rolling process to optics that the endless paper machine
provided for the newspaper printing press. Pynchon once wrote about
the protagonist of his novel, Slothrop, whose devout American ances-
tors were either the founders of a cult or paper manufacturers – and
paper is now the material basis for dollars, bathrooms, and Bibles,
that is, money, shit, and the Word of God. However, according to
Pynchon's analysis the same thing is now true of the power of the

media that historically superseded the word or book: it was a priest who bestowed roll film upon us. Celluloid itself had to be invented first, and this was accomplished by three Americans who wanted to use nitrocellulose to reduce the cost of the rather expensive and scarce elephant ivory used in billiard balls at that time. And nitrocellulose or so-called gunpowder must have been bestowed upon us even earlier than that; instead of the old monastic black powder, which I attempted to correlate with perspective itself, this was done by a Swiss chemist and an Austrian field marshal lieutenant named Franz von Uchatius, who will also be presented in the next lecture as the direct forefather of film technology. As the name already implies, gunpowder was not intended to serve cinematic or peaceful goals, but after its metamorphosis into the modern billiard ball Reverend Hannibal Goodwin, an enthusiastic amateur photographer who hated the heavy, unwieldy glass plates, was able to register his patent for roll film on May 2, 1887. The government's approval of the patent – along with a reimbursement of several million dollars – was not granted until 11 years later, though, because in the meantime the Eastman-Kodak company had already founded its billion-dollar fortune on the exploitation of Goodwin's patent.

We finally come to the era of high capitalism, and we are ready for the invention of film, which I will address in the next lecture. However, because I do not want to interrupt the technical connections again today, I would like to digress for a moment and outline a side of photography that is often left out: before the actual history of film I would like to examine the impact of photography as a storage medium on nineteenth-century culture and aesthetics.

3.1.3 Painting and Photography: A Battle for the Eyeballs

After 1836, there were two possible options available to everyone (and not only Napoleonic general staffs): either to write letters or books or else to send telegraphic signals. After 1839, there were also two options for images: either to paint or to photograph them. Arago's eulogy for Daguerre, which primarily emphasized the possible scientific applications of the first medium for storing images and completely denied that it would also exert competitive pressure on painters – particularly portrait painters since the arrival of the photographic portraits of Draper and Morse – was the understatement of the century. As everyone knows by now, the once massive business of painted portraits passed almost entirely over to photography, and under the competitive pressure of technology painters were only left with two options that essentially differentiated (to use Niklas Luhmann's term) them from photography. The first option was to change their style by no longer supplying image material from eyesight or from an old *camera obscura*, but rather from photographs. The long history of photo-realism – from Ingres (whose *La Source*, as I mentioned, was based on a nude photograph) to Gerhard Richter – is left to art historians. I will only say, as is clearly obvious in the work of Degas and his ballerinas, that the photograph as the new source material for painting replaced the imaginary (and therefore conventional painting's fixation on recognizing figures) with the reality of absolutely instantaneous contingencies and absolutely asymmetrical image fragments.

When viewed systematically, the problem with this first option is that it is only possible as long as other painters are unable to master the mimicry of certain photographic effects as well as those who introduced the new style. The other and historically more successful option for the painter was actually to differentiate between the artistic and the technical medium, and thus to only paint images that could not be photographed, such as images that do not represent any objects at all, but rather the act of painting itself. It is not necessary to emphasize that this option represented the mainstream of so-called modernist painting and historically it had practically no effects on everyday life.

With respect to everyday life, it is better to return to simple media like lithography, which will further our discussion of the politics of images, which is one of the leitmotifs of these lectures. In 1800, the romantic Novalis demanded – completely without reference to optotechnical media – that the as yet absolutist monarchy, and especially the Prussian monarchy here in Berlin, become more effective through reforms, and to become more effective it must arouse "belief and love" among its subjects. This eliminated baroque portraits that presented the prince's frightening yet also fascinating resplendence, as well as portraits that presented the prince in his lordly seclusion. Pictures of the solitary "Old Fritz" were immediately replaced with pictures of a king shown intimately or familiarly together with his wife, the proverbial queen Luise, as a married couple. And because the king's subjects had in the meantime also been remodelled into happy families, pictures of the ruler coincided with those of the ruled for the first time in history.

It was only later recognized that Flaubert, in his *Sentimental Education* – a novel about painters, dealers, and the revolution of 1848 – relates with all the cynicism available to him how the Prussian example became the accepted thing. According to Flaubert, King Louis Philippe (the very king to whom Daguerre owed his life-long pension and the Cross of the Legion of Honor) circulated countless lithographs on which he modestly appeared as a member of one of millions of bourgeois families:

> Then the talk came round to the dinner at Arnoux's.
> "The picture-dealer?" asked Sénécal [in the novel he is initially passionate about the revolution, but then he becomes its worst traitor]. "He's a fine fellow if you like."
> "What do you mean?" said Pellerin.
> Sénécal replied:
> "He's a man who makes money by political skulduggery."
> And he went on to talk about a well-known lithograph which showed the entire royal family engaged in edifying occupations: Louis-Philippe had a copy of the Code in his hand, the Queen a prayer-book; the Princesses were doing embroidery; the [young] Duc de Nemour was buckling on a sword; Monsieur de Joinville was showing his young brothers a map; and in the background could be seen a bed with two compartments. This picture, which was entitled "A Good Family," had been a source of delight to the middle classes, but the despair of the patriots. (Flaubert, 1964, p. 62)

It seems to me that this straightforward text needs no interpretation, but I would like to emphasize two points: first, it shows how effectively the politics of images functioned after the switch to infinitely reproducible and printable lithographs, and second it shows that media have repercussions on what they represent. In the mass medium of lithography, the royal family cast off all of its sovereign attributes and aligned itself with the mass of French bourgeois families.

As an automated form of lithography, photography only strengthened this trend. In Dessau, in one of the smallest German palaces, portrait photography caused members of the aristocracy to present themselves no longer in full dress with uniform and decorations, as they had for portrait painters, but rather they wanted to appear on their photographs wearing the simple black suits of normal citizens (Buddemeier, 1970, p. 86). Bourgeois realism was thus not only a style in literature and painting, but also in everyday life. From this, one can also infer the complementary need for media, which will compensate the aristocracy for this loss of face in the eyes of their secret bourgeois admirers. A first, harmless example are feature films like *Sissy*. A second example, which McLuhan brilliantly discovered, is the monocle, which became fashionable around 1850: an aristocrat would endure being photographed in the simple black suit of the nineteenth century, rather than portrayed in royal clothing; however, in revenge he wore a type of glasses that ensured even more brutally than the hand-made *camera obscura* in Brockes' *Rahmenschau* that in contrast to the monocle wearer, all others were reduced to the subjects of an optical media technology. Actors like Erich von Stroheim only needed to export or rather prostitute this trick of aristocratic German officers' eyes to Hollywood and a new film genre was born: the Nazi film (McLuhan, 1964, p. 170).

On the other hand, true bourgeoisation was implemented by the citizens themselves. The role played by photography (in contrast to Diderot's fictional warts) can be seen by looking at literary realism. Balzac, whose novels contributed the most towards the popularization of this new bourgeois realism, wrote in the foreword to *The Human Comedy* that his entire cycle of novels was like a daguerreotype of contemporary French society. As usual in the century of sciences, literature thus obtained its validation for the first time from a technical medium, which it could do more easily than painting because the medium appeared only as a metaphorical and not a real competitor.

As a metaphorical model, however, photography also appears to have had real effects on writing: Flaubert's equally magnificent and dismal *Madame Bovary* repeatedly mentions a *curé de plâtre* or plaster priest that initially stands intact in the Bovary's garden like a garden gnome, which is similarly mass-produced. It receives a few scratches during the first move, and finally, when the marriage collapses, it also falls to pieces. This priest does not serve the slightest function in the narrative other than to prove that the novel has not forgotten any visual detail within its fictional world – a forgetting that, in contrast to the realist Flaubert, actually principally befell his predecessors, such as Goethe and other classical writers. Against the backdrop of photography, literature therefore no longer simply produces inner pictures for the *camera obscura* that Hoffmann's solitary romantic readers became; rather, it begins to create objective and consistent visual leitmotifs that could later easily be filmed.

This does not mean that realistic writers (like painters) did not describe photography as a threat. The same Balzac who claimed to have drawn up all of his fictional figures like daguerreotypes also said to his friend Nadar, France's first and most famous portrait photographer, that he himself would dread being photographed. Balzac's mystical tendencies led him to conclude that every person consists of many optical layers – like an onion peel – and every daguerreotype captures and stores the outermost layer, thus removing it from the person being photographed. With the next photograph, the next layer is lost, and so on and so on until the subject disappears or becomes a disembodied ghost (see Nadar, 1899). Edgar Allan Poe, who also wrote about photography as one of the wonders of the world, made this phantasm universal by positing the thesis that images in general are deadly for their object (Poe, 1965, pp. 245–9). Poe's painter creates a portrait of his beloved without noticing that she grows increasingly pale the more that her oil painting acquires the color of human flesh. Painting, with its extensively discussed handicap of aging pigment, thus uses a photochemical effect against people as if it had become photography. As soon as Poe's fictional painting is completed, the painter's beloved drops dead. Media analysis can only emphasize once again how historical phantasms and collective symbols (as Jürgen Link would say) are directly based on technologies. The fears of Balzac or Poe merely illustrate the fact, as Arnheim theorized, that photography represents the emergence for the first time of a storage medium capable of reproducing the unimaginable materiality of the person being depicted.

This does not mean that the phantasm should be explained away, but rather just that it should be acknowledged as a fact, because the phantasmagorical fears of so-called humans – a category that according to Foucault was not invented until around 1800 as the subject of all possible fields of knowledge – had definite technological consequences. To begin with, all of the ghosts created using the *lanterna magica* since Schröpfer and Robertson, which Balzac eventually identified with the human itself, passed on to the new medium. A favorite pastime of occultism, which emerged around 1850 and which initially mimicked the electric telegraph, was to hunt for spirit photographs. The camera shutter was left open when there was nothing to be seen, such as in the dark, and these non-images were then developed in the Arnheim-like hope, so to speak, that a ghost invisible to the eye had materialized on the photographic film all by itself. This process was more successful than one could possibly imagine today, and it proves the half-truth of Adorno's dictum that spirits historically appeared at exactly the same time as the spirit of philosophical idealism was destroyed by media technologies. I say half-truth because the Counter-Reformation ghosts existing since Kircher's *lanterna magica* did not take a single syllable of Adorno into consideration.

Second, and this is the decisive point, what befell ghosts with so-called ecto-photography could also be applied to humans. An evil power, which was not always as fictional as in Chamisso's tale of poor Schlemihl, bought their shadows from them. As photography made advances towards snapshot photography, it became increasingly easier to capture people without their permission or even their knowledge. Balzac's nightmare proved well-founded in 1880 at the latest, when Alphonse Bertillon – whose brother, by the way, was notably the director of the statistical bureau of Paris at that time – became director of the Parisian police's identification bureau. Bertillon asked himself which visible and measurable features of a person remained constant for the remainder of one's life despite the changes caused by age or intentional disguise. He identified the following characteristics: the length of the head, the width of the head, the length of the middle finger, the length of the foot, the underarm length, and the length of the little finger. If the police were to store all these measurements on cards along with a careful description of race and any scars, birthmarks, and tattoos, the probability of mistaking the relevant individual for someone else would become virtually zero. Eleven measurements alone would already allow for 177,147 different combinations.

That is as far as we will go into statistics, and now we return to photography. In the age of literary realism, it was self-evident that Bertillon replaced these necessarily imprecise descriptions with photographs. From this time onwards, two portrait photographs were immediately taken of every arrested criminal or suspect, one from the front and the other in profile. Originally, the criminal even had to hold a ruler beside his face (Busch, 1995, p. 316). The criminal, with an example of the revolutionary international prototype meter in hand, thus measures himself. He is transformed into a test device of the new technique, just as Arago predicted. Even if this did not happen voluntarily, it was still an innovation. Photography, according to Arnheim's excellent definition, means that something real (whatever that may be, criminal or otherwise) leaves its traces in a storage medium. The user, in other words, no longer needed to create imprints, as was unavoidable under aesthetic conditions. Bertillon did not begin in a historical no man's land, but rather through the photographic recording of criminals he replaced older procedures such as the branding of criminals. In 1831, only eight years before Daguerre's invention, the last criminal was reportedly branded in Europe, while the search for new storage technologies had already started. Bertillon's criminalistics was an answer to exactly this search, as it replaced the arbitrary writing of the old European powers with modern scientific reading, which was far more efficient. The writing of brands, when they were read again a few years later, only gave the authorities the tautological information that someone had inscribed them, while the reading of criminal photographs, despite its apparent passivity, provided the authorities with non-redundant data about subjects for the first time. In a more realistic way even than Diderot's warts, which allegedly could not have been made up, criminal photographs produced identities.

As if this were not enough, Bertillon also successfully pressed for the international standardization of his so-called anthropometry, and he enforced this procedure not only against recidivist criminals, but against everyone: "It also proves useful in regard to law, forensic medicine, etc. as soon as it is a matter of allaying doubt about the identity of a person (legal identification cards, passports, personal descriptions on warrants, the identification of escaped lunatics, casualties, fished up corpses, and many others)" ("Bertillonsches System," 1905, p. 732). The model of the criminal nevertheless became part of our everyday life, which prompted Thomas Pynchon, who refuses every photo shoot and interview, to ask his readers: "Is that who you are, that vaguely criminal face on your ID card, its soul snatched by the government camera as the guillotine shutter fell?" (Pynchon, 1973, p. 134).

It must be pointed out again how right Manfred Frank is (although probably unintentionally) when he celebrates the human – freely quoting Schleiermacher and Sartre – as a collective individual. In the age of the monopoly of writing, Goethe could explain that individuals do not exist at all, but rather only *genera* or types. In spite of all attempts to project the protagonists of novels before the inner eye like a *lanterna magica*, no one knew what they actually looked like. Because language belongs to everyone, according to Hegel's insight, descriptions always already transform individuals into universals. The most glaring literary example of this transubstantiation was the arrest warrant that Georg Büchner inserted into his comedy *Leonce and Lena* after his own bitter experiences with the Hessian police in 1835:

FIRST POLICEMAN. Gentlemen, we are looking for someone, a subject, an individual, a person, a delinquent, an interrogatee, a rogue. (*To the* SECOND POLICEMAN.) Have a look, is either of them blushing?
SECOND POLICEMAN. Neither of them is blushing.
FIRST POLICEMAN. So we must try something else. Where is the "wanted" poster, the description, the certificate? (*The* SECOND POLICEMAN *takes a paper from his pocket and hands it to him.*) Scrutinise the subjects while I read: "A man . . ."

SECOND POLICEMAN. No good, there are two of them.
FIRST POLICEMAN. Numbskull! "... walks on two feet, has two
arms, also a mouth, a nose, two eyes, two ears. Distinguishing features: a highly dangerous individual."
SECOND POLICEMAN. That fits both of them. Shall I arrest them
both? (Büchner, 1987, p. 129)

So much for the comical aspect of old European warrants. Now for
the sad part: the actual warrant for the fugitive Georg Büchner that
the Grand Duchy of Hessen published in the *Frankfurter Journal* on
June 18, 1835 contained little more:

> Arrest warrant. Georg Büchner, student of medicine from Darmstadt,
> has fled the fatherland to evade criminal invesigation of his participa-
> tion in acts of treason against the state. Public authorities both at home
> and abroad therefore request that he be arrested and turned over to
> Councillor Georgi, the examining magistrate appointed by the Grand
> Ducal court of the province of Upper Hesse. Darmstadt, June 13, 1835.
> Personal description. Age: 21 years. Size: 6 feet, 9 inches according
> to new Hessian measurements. Hair: blonde. Forehead: very arched.
> Eyebrows: blonde. Eyes: grey. Nose: strong. Mouth: small. Beard:
> blonde. Chin: round. Face: oval. Complexion: fresh. Stature: strong,
> thin. Distinguishing features: near-sighted. (Büchner, 1985, p. 92)

In Büchner's ingenious simplification, the literary warrant reveals
how the high absolutist authorities had invented him very literally as
a single subject, which means that it conflated so to speak all actual
subjects into one subject. In the empirical press, like the *Frankfurter
Journal*, things were only marginally more complicated: Büchner's
warrant did not actually describe all the subjects of the Grand Duchy
of Hessen, but rather it only pertained to the ideal of a healthy, blond
student, and there were at least a few hundred real students who
fitted this literary model. Modern forensic evidence, on the other
hand, works with media rather than arts – its correlate, therefore,
is nothing but statistically singularized individuals who (as in Poe's
famous tale *The Man in the Crowd*) can themselves still be fished
out of the masses.

Two photographic examples related to modern forensic evidence
now follow. The first is fictional, and the second is its historical
confirmation.

The literary and therefore fictional example is taken from a book
by Gerhard Plumpe with the entitled *Der tote Blick* (The Dead Look).
In a slightly redundant way, this book shows how nineteenth-century

photography legally disrupted the copyright law concerning images (at that time, this of course meant hand-painted images), which had only just recently been introduced. Despite this preoccupation with laws (and not with media technologies), *Der tote Blick* contains several observations for which one can be grateful.

Plumpe summarizes the content of a comedy that the high pros-ecutor and forgotten poet Apollonius von Maltitz produced in 1865 under the title *Photography and Revenge*:

> The arrival of a traveling photographer at a bathing resort causes a disturbance, as the photographed guests feel that their portraits are an imposition in every respect. "That's how I should look?" – complains a young woman – "not simply ugly enough to horrify people, but mali-cious, like an ex-convict [. . .]. That is my poor deceased father's favor-ite child! Could you love these grimaces, my unfortunate Rudolph? – Everyone who looks at me this way must see me as the crooked daughter of a wealthy factory owner, who married an aristocrat only because of his money." And her mother, who is also photographed, is appalled: "Led to the altar from the nursery, beautifully named, deified by painters [. . .] sculpted in marble by Thorwaldsen – now in the hands of a charlatan." It continues this way for a while longer, and the confused indignation of the bathing guests is made complete when an "art expert" confirms the success of the portraits. "Do you find a faint similarity?" the expert is asked, and he answers: "It is not ideal, but rather perfect. It is not similar, but rather absolutely the same!" The horror of the guests eventually becomes so extreme that the doctor of the resort forbids the photographer from practicing his trade. The story takes its first turn of events when the photographer accidentally succeeds in photographing a criminal who is up to some mischief at the resort and is stealing from the guests. With the help of "realistic" photography, the thief is arrested. Photography is thus rehabilitated and the guests ultimately praise the mastery of his "art." (Plumpe, 1990, p. 193)

So much for the comedy *Photography and Revenge*. Mediocre poets can therefore also be good lawyers and even better media theorists. On the one hand, the media technique of photography destroys pre-cisely the "ideal" or imaginary, which sculptors or painters repro-duced again and again when they dutifully "deified" their models, because it manifests for the first time something real that makes even the noblest daughter suddenly look like an ex-con (guillotine grimaces, as Pynchon said). (For this reason, by the way, there has been a law in the German Reich since 1902 that gives every man and woman the "right to one's own image," which protects them against the misuse of photography.) On the other hand, however, photography also demonstrates the completely new ability to recognize and thus produce real convicts. And third, the fact that it did not become famous as "art" (in the old European understanding of the term) until or rather precisely after this criminalistic success proves – and here I deviate from Plumpe's thesis – that all talk about photography as art actually serves to conceal its strategic functions.

To conclude this digression concerning the cultural effects of the new medium, I would only like to point out that photography as legal retribution not only celebrated its triumph in literary fiction. As early as 1883, "a rapid photographic printing process" for arrest warrants, which Bertillon reportedly took over and standardized, enabled the arrest of "the dangerous anarchist Stellmacher" in Vienna (Eder, 1978, p. 441). As you may remember, nitrocellulose can be used to make either bombs, like the anarchists, or roll film, like the Viennese police. Between these two barrages, the anarchistic and the photographic, the human as collective individual explodes.

3.2   Film

3.2.1   Preludes

Now we come unceremoniously to the prehistory of film. For the storage of moving images it is just not enough to make a donnish assistant like Morse stand still for half an hour, as Draper did for the first photographic portrait, or to bind a criminal during an anthropometric sitting, as Bertillon did. Rather, it involves fixing the object or target precisely while it is fleeing and being able to reproduce this fixed movement again anywhere. For these two reasons, I will begin the prehistory of film with an American ship's boy named Samuel Colt. His history is actually slightly mythical and it will need to be expanded more correctly and precisely in a future lecture on the history of weapons technology, but it will suffice for today. In 1828, the ship's boy went to the East Indies and on the way he had a technical epiphany – namely, the revolver that is now named after him. Colt revolvers, as celebrated and not by accident in all western films, no longer aim their six shots from one man to another, but rather from one white man to six Indians or Mexicans at virtually the same time. This was the reason why Colt, whose factory almost bankrupted him, did not become a wealthy arms supplier until the American-Mexican War of 1847.

The colt not only introduced the innovation of being able to shoot six moving targets in quick succession at a time when contemporary soldiers still needed a full minute to load the next round in their muzzle-loaded rifles, but rather it fundamentally revolutionized the process of industrial manufacturing. For promotional purposes, Colonel Colt was always fond of demonstrating to his astonished visitors that it was possible to disassemble six colts on a table, jumble up their component parts, and in the end – despite this artificial introduction of statistics or noise – reassemble six fully functional colts once again from the individual pieces. I do not need to delve any further into the complicated prehistory of this trick, which can be traced from Ludwig XVI's artillery to the almost forgotten but not unimportant British-American War of 1812 and the US Army Ordnance Office. It is enough to say that even though Colt did not invent the principle of the industrial serial production of interchangeable parts, he still publicized it very successfully. The series of shots in time and the series of devices in space were two equally important aspects of one single innovation. The arms supplier of Napoleon's great army worked on similar standardizations at practically the same time, as did the English computer pioneer Charles Babbage, particularly with regard to screws and other precision mechanics. As you know, however, Colt's model prevailed in America – and this was actually for the simple reason that every conceivable emigrant with every conceivable occupation that was not demanded elsewhere streamed into the country of unlimited serialism. There were only two groups of workers who did not emigrate from Europe, where they enjoyed much better working conditions: skilled labor and the military. And behold: the manufacturing technology of Colt's revolver compensated for the first shortage and the weapons technology compensated for the second.

Both of these aspects were also crucial for film. First, concerning the seriality of the production process, film distinguishes itself from photography in that the sender's finished product – the film in reels – is entirely useless if a projector with precisely the same specifications is not available on the receiver side. The purchaser of a photograph does not himself need a camera, but the purchaser of a film needs a projection room and a projection device. While Shannon's channel concept is rather anachronistic and unsuitable for photography, as I have said, film comes considerably closer to this concept and thus requires highly industrial conditions. It is no coincidence that many early film producers came from the sphere of precision engineering (Faulstich, 1979, p. 159).

The seriality of shooting a revolver, on the other hand, naturally corresponds to the serial time in film, into which the movements of the filmed object must be broken down. In terms of pure mathematics, this has not been a problem since Aristotle's theory of movement was adopted in the early modern period. In the fourteenth century, as I have mentioned, Nicolas Oresme already sketched the individual phases of the flight of a missile on paper, and Leibniz developed differential calculus around 1690 in order to calculate the ballistics of cannonballs. dy over dt means analyzing the results of an arbitrary mathematical function in extremely small intervals of time t, and these intervals eventually approach zero until the differential quotient indicates the tangent and that means the change of the relevant function itself at all individual points in time.

Technically, however, this border crossing is simply impossible because (according to Shannon) there are no infinite scanning speeds. It was thus replaced with the problem of how small the segments of time must be made in order to provide at least the appearance of such a border crossing. At the same time that Charles Babbage constructed his first proto-computer, which converted Leibniz' differential equations into technically realizable difference equations, the nineteenth century developed a machine that operated even below the smallest difference that would still be physiologically perceptible. But that suddenly changed the technical question into a physiological question and the construction of machines thus changed into the measurement of human senses.

To identify this new physiology of the senses, it will suffice first of all to point out in general that its scientific structure would have been inconceivable prior to the nineteenth century. In his remarkable book about the techniques of the observer, Jonathan Crary even postulated the thesis (inspired by Foucault's historiography) that the turn away from physically natural optics, as represented by Lambert, for example, towards physiologically embodied optics was a veritable scientific paradigm shift. The principle support for Crary's thesis is no less than Goethe, whose theory of colors was fundamentally based on the phenomenon of optical after-images. Someone looks at something red for a few minutes, then closes the eyes – and suddenly the complementary color green appears to these closed eyes. Goethe boldly concluded from this, as I already mentioned at the very beginning of these lectures, that the eye is like the sun: out of its own creative activity it generates a suitable complement to every passively pre-existing color, and the end sum is always a totality.

Crary's thesis reduces many events in the history of science that led to photography and film to a brilliant denominator. Nevertheless, I would like to raise two objections. The first concerns Crary's over-emphasis on the body, which is fashionable among contemporary scholars. There seem to be entire branches of scholarship today that believe they have not said anything at all if they have not said the word "body" a hundred times. There is no doubt that in the nineteenth century the geometric model of optics, which prevailed from the time of Brunelleschi to Lambert, was replaced with a materialistic one, but that by no way means that the material effects of light always impact on human bodies and eyes. It can just as easily be, as we have seen, Schulze's photochemical effect on silver salts and, even more conclusively, Herschel and Ritter's history of infrared and ultraviolet. Crary's thesis would therefore be more precise if he had not spoken about physiology but rather about material effects in general, which can impact on human bodies just as well as on technical storage media.

Second, I do not see how Crary can equate Goethe's gentle experiments with the more brutal and in my eyes first true physiological experiments and self-experiments of his successors. Goethe himself boasted of his "delicate empiricism," and he surely never caused pain for the sake of his theory of colors. However, the Weber brothers, to whom the sciences of motion (as they were called in the nineteenth century) owe much, falsified the alleged creative power of Goethe's eye by simply delivering a mechanical blow to their own eyes: what then emerged as an after-image or lighting on the retina was no longer a totality, but rather the trace of a shock (Crary, 1991).

The Leipzig scientist Gustav Theodor Fechner was even worse than the Webers because he first attempted to prove Goethe's precious theory of after-images experimentally. As a physicist, Fechner also wanted to determine the measurable quantities and measurable periods of this after-image effect, and he spent three years reading all the relevant books on the subject and then staring into the sun. At the end of this series of experiments, which exposed his eyes to two rather opposed extremes, he was blind and fit only for a mental institution (see Lasswitz, 1910). You can see that in the nineteenth century the physiology of the senses did not simply ruin experimental rabbits – or rats, like today – but rather it ruined the research pioneers themselves. Media always presuppose disabilities, and thus optical media also presuppose the blindness of their researchers (in addition to a lack of natural pigments). Enlightenment philosophers like Diderot or Condorcet had only postulated theories about the

148

blindness of others, because the Enlightenment itself was supposed to be pure light. Fechner, on the other hand, was able to write the general mathematical formula of all sensory perception, the so-called basic law of psychophysics, precisely because he sacrificed his eyes to research his subject and then only managed to improve his condition again through sheer force of will. According to this basic law, a linear increase in objective stimulation only corresponds to a logarithmic increase in subjective sensation; by the same token, an exponential increase in stimulation is necessary for a linear increase in sensation – in Fechner's tragic case, therefore, the sun must shine four times brighter to blind twice as much. With such optimistic and also not undisputed assumptions about sensory resistance, one can imagine how much solar power Fechner exposed his eyes to.

Fechner is admittedly less important for the physiology of film than another blind man, to whom we will shortly come. Fechner only serves to illustrate a research field that began making numerical statements about perceptual processes and above all stimulus thresholds. It is clear that eyes can only believe in the apparent continuity of film movements when the projected images change quickly enough that the sequence of individual frames drops below a certain temporal threshold. The so-called positive after-image then takes effect. In contrast to Goethe's celebrated concept of the negative after-image, the positive after-image occurs when the eye continues to see an object in the same place a moment after it has already disappeared or moved away. This happens because the stimulation of the nerve fibers only wears off gradually, and the after-image remains in the same color as the original image rather than the complementary color, as with a negative after-image. Since about 1750, it has been known (or rather rediscovered, as Ptolemy's *Optics* was reportedly aware of the after-image effect) that the positive after-image lasts for an eighth of a second. The eye is therefore no longer able to differentiate movements faster than this from one another. It was not until the nineteenth century, however, that researchers proceeded to take advantage of this effect with small technical devices that produced illusionistic effects as toys. In 1824, for example, Sir John Herschel, the son of the aforementioned astronomer and discoverer of infrared, rotated a coin so quickly that by all appearances the front and the back, the number and the emblem, were visible at the same time as a single image (Zglinicki, 1979, p. 109).

And yet the after-image effect alone is still not enough to make cinema possible. It only supports the cinematic illusion in one respect: it dampens the flickering during the film advance and completely suppresses it upon reaching the flicker fusion threshold. But to produce the illusion that one and the same object has moved from the place it occupied on frame A to another place on frame B, another optical effect must be added: the stroboscope effect. Hopefully, it is not necessary to say much about the effectiveness of this effect, as all of the better discos employ stroboscopic lights so that people's dance movements can be cut up into their individual phases, much like film editing. The twentieth century, in other words, has successfully reimported a film effect into everyday life. The nineteenth century, on the other hand, had to first discover the stroboscope effect to make film possible at all.

It is a great pleasure to inform you that the great physicist Michael Faraday was among these discoverers. Faraday will appear later in these lectures as a genius at theater lighting, but in 1831 he also discovered electromagnetic induction or the possibility of producing voltage and ultimately alternating current through the circular motion of an electric circuit in a magnetic field. What he discovered for optics is not so very far from induction, because it already prepared for the possibility of one day electrifying optical media like film and television. As in the case of the rotation press or the revolver, circular motion once again plays the decisive role, which will culminate in roll film. Through his fundamental electromagnetic discovery, Faraday took notice of circular motion in general, and he reportedly observed two gears in a mine whose motion was normally not perceptible at all because of speed and thus because of the after-image effect (Zglinicki, 1979, p. 114). On the basis of this observation, Faraday constructed purely experimental gear couplings, until he determined a new optical law: the periodic breaks in the equally periodic images – which occur approximately when the front gear allows the viewer to see the individual teeth of the rear wheel but then conceals them again – leads to the lovely illusion that the eye mistakenly identifies tooth A from image 1 with another tooth B from image 2 with a third tooth C from image 3, etc. A virtual movement thus emerges, and at certain rotation speeds or frequencies the gears even virtually stop. Electro-technicians and information theorists like Shannon would say that the sampling frequency together with the frequency of the samples produces an aliasing effect, which is perhaps a free English translation of Brecht's alienation effect. The possibility of this aliasing effect is only present when the sampling frequency is not at least twice as large as the maximum frequency in the signal of interest. For this reason, sophisticated filter chains provide for a meticulously precise observation of Shannon's sampling theorem during the digital recording and playback of compact discs. And the fact that the stroboscope effect does not hinder but is actually necessary for film says everything about the difference between film and electro-acoustics, the imaginary and the real (this is already a slight anticipation of statements that are yet to come). It only becomes obtrusive and disruptive when film scenes themselves demonstrate the very effect on which they are based. You all know that when a western is shown at 24 frames per second and the famous covered wagons of the American pioneers have exactly the right frequency, their spokes appear to be standing still or even running backwards.

So much for Faraday, who admittedly appears to have been more interested in a basic theory of frequency than in its media-technical applications. The physicist neglected to demonstrate his stroboscope effect not only with the teeth of gears and slits but rather with images, which was a small but decisive step in the development of film. However, Joseph Plateau, the Belgian professor of experimental physics and astronomy at Ghent University, was working on optical illusions at the same time and completely independent of Faraday. In 1832, he thought of feeding the stroboscope with 16 drawings of a dancer, presenting her in successive phases of movement and ending once again in the initial position.

On the outer edge of the disk and between the individual images there were 16 slits. When the spectator positioned the disk in front of a mirror, set it in motion, and continued looking through the same slit into the mirror, the dancer herself would proceed to perform endless pirouettes or circular movements. For the first time, a technical trick had changed the zero frequency, which was the rate at which all representative artworks had been displayed ever since the Stone Age, into frequencies as high as one likes. This must have so deeply fascinated Plateau that he was no longer able to leave his optical experiment alone and he gradually went blind; in contrast to his colleague Fechner, however, his blindness was permanent.

Perhaps the extent of the sacrifice that Plateau made enables us to appreciate the advance that his stroboscope represented. If one thinks back to Athanasius Kircher's smicroscope, which was able to present the 14 Stations of the Cross one after another, the first important difference is the novelty of Plateau's representation of the successive phases of the dancer's movements. The 14 Stations of the Cross were 14 different images as such, one on the Mount of Olives, one with Pilate, one on Golgatha, etc., from the night before Good Friday until the famous sixth hour. The 16 drawings of the dancer, on the other hand, are absolute snapshots of one and the same object – and this was done seven years before Daguerre was able to reduce photographic recording time to two to three minutes. Imagine if the devout Jesuit Kircher had presented the passion play as endless pirouettes. With his virtual circular motion, Plateau is a worthy contemporary of all the acoustic experimenters between 1830 and 1880, or between Weber and Edison, who made analogous attemps to achieve millisecond recordings of sound and speech periods, which finally culminated in Edison's phonograph in 1877.

That means at the same time that film was not yet technically possible during Plateau's lifetime simply because instantaneous photography lagged far behind the rotation speeds that were attainable with the stroboscope. For this reason, only scientific devices and toys were initially developed from the stroboscope. The device was suggested by Doppler, who also discovered the acoustic effect that was named after him: to analyze motion, whose speed strips it of every visual perception, Doppler employed a systematic reversal of the stroboscope, which did not set images and slits into periodic motion but rather the light source itself – as a rapid succession of electrical sparks, for example (Zglinicki, 1979, p. 120). This is precisely what in the meantime is employed in discos, most likely to train the speed of our perception – in defiance of all physiology – for the extreme requirements of a technical war.

The development of toys also proceeds in a similarly militaristic way. You may recall that one of the reasons why Leibniz developed his wonderful differential calculus was to make missile trajectories calculable. Now in 1811, a certain Franz von Uchatius came into the world, which at that time still had an Austrian Empire. In 1829, Uchatius voluntarily joined the second field artillery regiment as an artillery gunner. After graduating from the Bombardier Corps Academy in 1837, he was promoted to sergeant and commissioned to teach physical chemistry. In 1841, he began his – at least for the Austrian artillery – groundbreaking research on cannon casting, which ultimately led him to the invention of steelbronze and also of the aforementioned explosive Uchatius powder, which is chemically closely related to old roll film. As a reward, Kaiser Franz Joseph promoted him to Field Marshal Lieutenant and at the same time, because this promotion would have otherwise been impossible, made him a baron. In the symbolic world, on which monarchies are based, things fell into place very well, but Austria-Hungary did not have the least interest in the technical real world. After walking a great distance, a certain Mitterhofer from South Tyrol was permitted to present his wooden typewriter – the very first that we know of – to this same

152

Kaiser. He received a personal donation, but his machine did not go into production. Uchatius fared even worse: due to bureaucratic difficulties with the introduction of Uchatius cannons in the imperial army, the artilleryman and Field Marshal Lieutenant became his own target. On June 4, 1881 he shot himself in the head (Zglinicki, 1979, pp. 130–5).

After this requiem we now return to the history of film. Before Field Marshal Lieutenant Franz von Uchatius took aim at himself, it must have been important to him to teach all cadets and officer candidates the principle of artilleristic shooting. Plateau's newly discovered stroboscope lent itself easily to this purpose. Like Edison's kinetoscope, it also admittedly had the one crucial disadvantage that it was not a mass medium, but rather it always only allowed a single viewer to look through the observation slit. For his lectures on weapons technology, therefore, Uchatius spent his scarce free time first combining the well-known *lanterna magica* with the stroboscope. And behold: all of the cadets were immediately able to watch Uchatius' sketches of projectiles flying through the air at the same time, as they were projected onto the wall of the auditorium. The *lanterna magica* thus no longer produced merely virtual motion, like Schröpfer's curtains of smoke, but by means of a crank – much like early cinema – made successive images really dance in front of a fixed light source. It is no wonder, then, that Uchatius' constructions were purchased by a carnival showman, who gave magical performances that turned the weapon back into money, as Schröpfer or Robertson had once done (Zglinicki, 1979, p. 133).

What is interesting here is not this reversion, but rather the fact that the individual technical elements of film – the recording device, the storage medium, the projection apparatus – were combined with one another very gradually and in stages. Technical media are never the inventions of individual geniuses, but rather they are a chain of assemblages that are sometimes shot down and that sometimes crystallize (to quote Stendhal). After Uchatius combined the stroboscope and the *lanterna magica*, the only element that was still missing was the *camera obscura* that Talbot had already automated. However, its inclusion in the communication system known as film not only encountered technical obstacles, such as exposure times that were much too long, but also the traditionalism of an entire artistic epoch. As contemporaries of these experiments, draughtsmen like Toepfer in Geneva and Wilhelm Busch in his village near Göttingen were actually quick enough to grasp the new method of drawing in phases as an art form, which means quite simply that they invented the comic strip, which led in turn to the animated cartoon. However, it was precisely because painters learned to break motion down into successive phases in the nineteenth century that they did not want to relinquish this new art form once again and replace it with technical media. On December 1, 1888, a certain Emile Reynaud received the French patent number 194 482 for his projection praxinoscope. As the name already suggests, this device projects moving stories, which Reynaud drew on perforated and flexible ribbons as animated films. All of these features, but above all the interplay between the perforations and a gripping mechanism, guaranteed perfect synchronization. The musical accompaniment synchronized with the projection also practically anticipated Edison's kinetoscope, except that throughout his lifetime Reynaud stubbornly refused to replace his "artistic" drawings with photographs (Zglinicki, 1979, p. 136). As Hölderlin so accurately wrote, it is hard to leave a place when one lives near the source. And what lives closer to the source than art according to European tradition?

3.2.2 Implementation

It was thus left to a European emigrant to the US to take the last step. Edward Muggeridge from Kingston-on-Thames, who changed his name to Eadweard Muybridge either out of Anglo-Saxon pride or an American desire for self-promotion, combined for the first time all the elements of film: instantaneous photography, *lanterna magica*, and stroboscope. Muybridge's zoopraxiscope of 1879 showed, as its name claims, life (Zglinicki, 1979, p. 175).

The key word "life" naturally compels us to disregard momentarily this lecture's physical image of the world and to touch upon nineteenth-century zoology. However, to understand Muybridge's feat we are now in the same fortunate position as film history: we only need to combine the already existing or already recounted elements. Colt's revolver, Daguerre's photography, Weber's acoustic experiments – all of these elements recur once again.

The initial push was supplied by acoustics. It was possible to record the frequencies of the human voice long before Edison, but there were no researchers who also thought to play back this voice at another place and time. The trick simply consisted in sending the voice into an amplifying funnel at whose end a membrane vibrated. The other side of this membrane was attached to an innocent hog's bristle, which finally scrawled the captured frequencies onto a lampblackened glass plate – provided that the experimenter rolled this

154

plate past the bristle fast enough. In this simple way, for example, the very voice of the British phonetician who was the inspiration for Professor Higgins in Shaw's *Pygmalion* and the musical *My Fair Lady* has been preserved to this very day. For the first time, the physiology of a living human being was coupled with a storage medium rather than a chain of symbols with a repertoire of signs, as with writing.

3.2.2.1 Marey and Muybridge

Phonographic recording, which at that time was also called visible speech, became the accepted thing among European physiologists. When France established a professorship in natural history at the Parisian Collège de France at the instigation of the great physiologist Claude Bernard, for whom we have Zola's entire naturalism to thank, the holder, Prof. Étienne-Jules Marey, immediately began constructing devices. Doctors like Goethe's Mephisto had actually already strongly recommended taking the pulse of, above all, women with tenderness, and it was precisely because of such pleasures that it had not occurred to a doctor before Marey to replace his own hand with a machine. At the Collège de France, one device emerged after another: a heart recorder, a pulse recorder, and finally also a device that was connected to the four extremities of animals and could record their movements. None of these devices bore the least similarity to photographic cameras, but rather they worked, exactly like visible speech, with a pencil and a steadily moving paper cylinder.

As only luck would have it, Marey became acquainted with a captain in the French army who was also a horse enthusiast. This captain converted the results of the professor's measurements back into traditional art. His visual reconstruction of the measurements of a horse's legs manifested the incredible fact that there is a moment while galloping when only one of the horse's legs is touching the ground.[7] The Anglo-Saxon world in particular was overrun with watercolors featuring horses and riders, yet there was not a single picture showing the leg position that Marey claimed.

Imagine for the blink of an eye if film had been invented in India and the leg position was not that of horses but rather of women and men according to the rules of the Kama Sutra . . .

---

[7] Kittler is incorrect here and in the following paragraphs when he states that one horse's leg is always touching the ground during a gallop. The theory of "unsupported transit" referred to here actually claimed that all of the horse's legs were in the air at the same time.

Instead, the story continues in puritanical America, where Leland Stanford Sr. became a millionaire by constructing a pacific railroad and consequently, long before Ronald Reagan, also became governor of the state of California. Stanford, the horse enthusiast and trotting horse breeder, saw the measurements of the horse's legs, but was unable to believe the results of Marey's experiments. We can only assume the reason was that the watercolor images of horses' legs remained so deeply imbedded in the subconscious minds of people who had not yet seen a film. However, where belief is lacking, in America there is money and experimentation. Stanford had the fortunate idea of calling in Muybridge, the landscape photographer, and commissioning a photographic test of the horse's leg problem.

Muybridge thus exchanged the Californian wilderness for Californian civilization, or the timelessness of Yosemite Valley, which he had immortalized in his landscape photographs, for the millisecond realm of telegraphy. Stanford's breeding establishment for trotting horses was located at his ranch in Palo Alto, precisely where the Leland Stanford Junior University – renowned for being the best for the study of electronics – stands today. Muybridge constructed a white wall with a short race track in front of it, and in front of this race track he placed a row of 12 instant cameras, which were all connected electronically. With relay circuits supplied notably by the San Francisco Telegraph Supply Company, another media industry firm, Muybridge succeeded in triggering these 12 cameras one after another at intervals of only 40 milliseconds, whereby each individual camera had a shutter speed of a single millisecond. Then all he needed was to make a horse gallop along the race track and Governor Stanford had a black-and-white photograph proving that during a certain phase of movement while galloping only one of the horse's hooves was touching the ground.

You see: Muybridge's experimental set-up no longer had even the smallest resemblance to the stroboscope, but rather it recorded movements for as long and extensively as the experimenter wanted and the race track allowed. Cylindrical storage media, which – from Plateau to Marey – were confined to repetitions and periods, and thus choreography and poetry, were superseded by the prose of science and later also of entertainment media. You know that in poetry, which was formerly identical to dance, everything must come back around as in the stroboscope; in novels, on the other hand, there is always an unforeseen and contingent future, as in Muybridge's series of instantaneous photographs, which can consequently also only stop through an interruption. All of so-called modern life therefore depends entirely on nineteenth-century media technologies.

Muybridge's empirical confirmation of Marey's experiment took place in 1878. Five years later, the very same instantaneous recording technology used by Muybridge also made all of Field Marshal Lieutenant von Uchatius' dreams of artillery pedagogy come true. The physicist Ernst Mach sealed a wire in a glass tube, connected the wire electrically to the shutter of an instant camera, and then fired on the glass tube. The result was a photographic speed record: a real bullet, rather than a drawn one as in the stroboscope, generated shock waves as it entered the chamber. It is no wonder that the same Mach was also the first to achieve the opposite record: the time-lapse film for analyzing infinitely slow movements (Eder, 1978, p. 523).

Muybridge, as far as I can trust my optical memory, also invented another trick. The library of Stanford University, which the old governor founded on the site of his former horse breeding establishment to memorialize his son (whose death was caused by bungling European doctors, of course) and to prevent similar tragic cases in the future through science, contains stacks of enormous photo albums, in which Muybridge gradually shifts from horses and cows to people. Whenever an athletic Stanford student and sprinter enters the photograph from behind, he is naked, but whenever he turns his front towards the camera a swimsuit suddenly appears out of nowhere, as if the image has been retouched. Muybridge thus invented the stop trick, one of the most important film tricks, long before George Méliès. As far as I can see, frontal nudity first appears in Muybridge's photographs only during his later time at the University of Pennsylvania.

The purpose of the swimsuit in California is clear: only the nude photographs require some explanation. After Muybridge finally abolished the hand of the artist (which had appeared so irreplaceable to his predecessor Reynaud) and manufactured a pure multimedia system, he also wanted to reform painting. His magnificent volumes on "animal locomotion" were published for the express purpose of preventing artists from drawing or painting false positions, like the galloping horse. Muybridge's nude photographs provided them instead with a scientific model of all possible body movements. Like Renaissance perspective and the *camera obscura*, instantaneous photography was supposed to discipline art. Admittedly, this time it did not involve abolishing the dominance of the symbolic, as in medieval holy images, and introducing a human scale through perspective, but rather at the end of the nineteenth century the imaginary also had to believe in it. The reason we never see three legs of a horse in the air is because the eye projects a familiar general shape on all of the phases of an animal's movements.

I am pleased to be able to say that Muybridge's propaganda was a crowning success in at least one famous case. During his European travels, Muybridge's patron Leland Stanford Senior also went to Paris, where he became acquainted – presumably in Muybridge's company – with the painter Meissonier. Jean Louis Ernest Meissonier, who unfortunately hardly anyone appreciates any more except for Salvador Dali and myself, had already painted practically all of the legs of the horses in Napoleon's Great Army on extremely expensive canvas out of an explicit admiration for Napoleon's "organizational genius" (Gréard, 1897, p. 47), but he confessed that he only knew how to represent these legs while striding and trotting and not while galloping (Gréard, 1897, p. 194). Stanford demanded that Meissonier paint his portrait, which the painter would have refused to do had Stanford not revealed to him on this occasion the secret of Muybridge's instantaneous photographs of horses. Meissonier was converted, and it was because of him that this secret was eventually conveyed to the vice-president of the Paris Academy of Fine Arts. From then on, Meissonier employed photographic source material to paint real horses' legs rather than picturesque imaginary ones. According to his biography, he used these photographic models "only for verification," but in his private park in Poissy, near Versailles, he built a short railroad track, seated himself in a sleigh-like locomotive whose speed could be adjusted, and studied a galloping horse in motion with his own mobilized painter's eye (Gréard, 1897, p. 73). You see once again how the railroad replaced the horse in media history, and how the nineteenth-century railroad journey celebrated by Schivelbusch (1986) was not limited to being passively transported, but rather in the wonderful case of Meissonier it was already an active automobile in the literal sense of a forward-moving, self-propelling technology. It is hard to say whether this tremendous expense made Meissonier's war paintings more valuable or realistic, and there is no need to know. The death of traditional painting was quickly approaching, for in the same Paris salon where Stanford had met Meissonier, Muybridge also met Marey.

For the first time in this history of invention, therefore, we are confronted with a case of positive feedback. Sixty years earlier, Niépce and Daguerre had met purely by chance, even though they both lived in France, and their collaboration had to be insured through a civil contract. Now, in 1882, Marey's machine for measuring movement inspired Muybridge to design a follow-up invention in California, and Muybridge's serial photography, in turn, inspired Marey to design his own follow-up invention. Muybridge, as I have said, went as far as inserting his sequences of images – which were photographic and no longer merely drawn – into a stroboscope in order to be able to project real movements. This was still not a film, however, not only because a minute of horse galloping would have required 270 images instead of the available 12 (see Clark, 1977), but also because Muybridge was stubborn: as a prior landscape photographer, he never gave up using the heavy, immovable glass plates that Niépce's cousin had introduced to photography in 1847. For this reason, namely, because a human life is far too short to comprehend avalanches of technical innovations, teamwork and feedback loops become essential . . .

Muybridge and Marey, these twins or Dioscuri who were present at the birth of film, were both born in 1830 and died in 1904. It is therefore possible to shift from one to the other without any problems. After the soirée at Meissonier's, Marey realized that all of his machines for measuring heartbeats, pulse rates, and the movements of horses' legs had been historically superseded by Muybridge's serial photography – with one exception. By holding tight to the unifying, linearizing power of writing paper, Marey always only needed one single piece of equipment, while Muybridge had to position 12 different cameras. The task, therefore, was to dispose of 11 cameras and still be able to supply serial photographs. In the process, Colt's good old revolver was once again honored, as it had also reduced the need for six pistols down to one. In 1874, a French astronomer, Pierre Jules César Janssen, had already converted the revolver from the wars with the Red Indians into a revolver for the stars in order to capture 18 different positions of the planet Venus on a single photographic plate, whereby the astronomical revolver repeatedly closed his camera lens between these 18 instantaneous recordings with a Maltese cross (Zglinicki, 1979, p. 170). While Marey still had to install his individual images into a stroboscope by hand, this new device supplied the stroboscopes all by itself. After this preparatory work, it was easy for Marey to improve on Jannssen's astronomical revolver. Marey developed a device that was roughly 50 centimeters long, which he dubbed a chronophotographic gun or *fusil chronophotographique* because it handles photographs much in the same way as a gun. It was braced against the shoulder for stability, it had gun sights for aiming, and firing the trigger produced an instantaneous photograph, whereupon the barrel, which contained 11 more unexposed negatives, turned 30 degrees, bringing the next negative into position. If the metaphor of shooting a photograph was ever taken literally, then it was in the case of Marey, whose assistant

Georges Demeny was reportedly even ordered by the French general staff to record and optimize the standardized marches of soldiers using serial photography in 1904.

At a time when France was mourning a lost war and the lost provinces of Alsace-Lorraine, Marey's photographic rearmament considerably helped his career. The physiologist, who had already received his own physiological institute, rose even further to become president of the French photographic society. In this position, he put into place the penultimate step to film technology. The rigid barrel or disk, which had been maintained from Plateau's stroboscope through to the chronophotographic gun, was transformed into a flexible roll, which was transported automatically past the lens through a clock-work mechanism in the camera. And even though it still was not a celluloid roll, the technique of film recording was in principle fixed. Marey would only have had to put his photographic paper rolls into a projection apparatus driven by clockwork, but he did not, and he thus missed the chance of becoming Lumière or technical light itself. For this reason, we are still faced with the pressing task of recon-structing the commercialization of the half-military, half-scientific technology of instantaneous photography through Edison and the Lumière brothers.

3.2.3   Silent Film

We no longer need, as in previous lectures, to represent the history of this industrialization as a detailed account of individual inventions. Up until now, the presentation of these individual inventions illus-trated the simplest, namely earliest attempts to solve the fundamental problems of optical media technology. This no longer applies, as the individual inventions and patents related to film began to explode in 1890: while there were only around 200 film-related patents issued worldwide between 1875 and 1890, this number had already risen to 500 between 1890 and 1910. In the age of industry, therefore, film emerged from pure teamwork. That is why I will only make rough cuts that treat silent film, sound film and color film as epochal structures. The two world wars will also serve as crude landmarks.

Before discussing the actual development of silent film, I want to point out again how far Marey had already come conceptually despite the fact that he failed to project his sequences of images. In 1891, his assistant, the aforementioned Georges Demeny, developed the won-drous photography of speech or *photographie de la parole*. Precisely like the earlier telephone inventor Alexander Graham Bell, Demeny

160

also wanted to cure a physiological handicap, namely deaf-muteness, through media technology: he combined a photographic gun *à la Marey* with one of Edison's phonographs, turned the experimental apparatus towards himself, as was usual at the time, and shouted two very significant sentences into both devices at the same time: first *"Vi ve la Fran ce!"* and second *"Je vous ai me."* It was the first declaration of love in history that was no longer directed at women, but rather at the medium of film, which has since become the norm. At the same time, however, it was also a lesson for deaf-mutes, whose mouths reportedly managed to produce audible declarations of love to no one at all by imitating Demeny's individual mouth positions.

We can see, then, how desperately the coupling of optical and acoustic media technology was sought from the very beginning, long before the introduction of sound film. This is even truer of the inventor who actually applied it to commercial cinema: Thomas Alva Edison himself.

Edison worked as a telegraph operator during the American Civil War, which made him partially deaf but also provided him with technical know-how and money. On this double basis, his laboratory – which was truly the first laboratory in the history of technology – first realized two dreams of the century: the mechanical recording of sound, namely the phonograph, and the perfect light source, namely the light bulb. The phonograph, as the first form of visible speech capable of also being played back, was developed in 1877 as a byproduct of Edison's attempt to accelerate the transmission of telegraph signals. It therefore shows that the opposition between discrete and analog media was already beginning to become fluid in Edison's time. The light bulb, which in turn led to the tubes that were the basis of all electronics for a long time, emerged from the search for a light source that would avoid the smoke and fumes (and thus the signal noise) of ancient candles. At the same time, it was also supposed to avoid the short life span of carbon arc lamps and the rather deadly dangers of gaslight – the two light sources that had immediately preceded the electric light bulb in the nineteenth century.

It is important to note here that Edison's kinetoscope – the immediate predecessor of film – was directly connected to these two previous inventions. To begin with, this is true biographically. The phonograph and the light bulb made Edison renowned, so to speak, for being able to invent as if on command. One of the first to bow to his fame was the great Berlin physicist Hermann Ludwig Ferdinand von Helmholtz, the founding hero of all eye and ear physiology. His acquaintance with Muybridge provided Edison with the same fame in America. During his trip to France in 1881, where his attempts to meet the science-fiction novelist Villiers de L'Isle-Adam unfortunately fell through, Edison also made the acquaintance precisely not of poets but rather of fellow researchers like Marey. After both of these meetings, nothing seemed more obvious to Edison than turning a scientific experiment into a money-making entertainment medium.

After the phonograph and the light bulb, therefore, Edison also developed the first commercial film system. And because entertainment media had to be sold and distributed worldwide, Edison's stroke of genius was standardizing the serial instantaneous photography of Muybridge and Marey, just as Colonel Colt had standardized the revolver as the first serial murder weapon. After he had become acquainted with Marey in Paris, Edison found Eastman-Kodak's celluloid film. By choosing the 35 mm format and furnishing the film roll with perforations, which have remained the standard practically ever since, Edison solved all of Marey's problems of film synchronization in one fell swoop. He subsequently constructed a component called the kinetograph, which recorded moving pictures, as well as a compatible or standardized component called the kinetoscope, which could play back the developed film.

One year later, Edison finally acquired the patent for the so-called escapement disc mechanism from another American, which ensured that the individual frames of the film stood beautifully still during the sixteenth of a second in which they were recorded or observed, while all further transport between the individual frames fell precisely in the pauses in between. Since this fundamental solution, at the very latest, film has been a hybrid medium that combines analog or continuous single frames with a discontinuous or discrete image sequence; this will be amplified even further in connection with television. In 1888, Edison placed this entire digital-analog construction in a box that was essentially an electrified version of the peep show cabinets at eighteenth-century fairs: an electric motor pulled the film roll, which was illuminated from behind by a light bulb, past a magnifying glass, through which the primarily individual observer, upon inserting a coin, was supposed to follow the moving film and experience the illusion of continuous motion. Edison's sales success was so great that nickelodeons, as they were called, sprang up everywhere in America (they are predecessors, so to speak, of contemporary arcades). William Fox, among others, later made his money as the inventor of Movietone talking newsreels (Zglinicki, 1979, p. 208).

In addition to the financial effect of this new illusion, it is also important to point out its technical basis: the acoustics of telegraph signals also provided apparent continuity on the basis of actual discontinuity, which was no longer humanly controllable, and this had originally inspired Edison's phonograph. For this reason, he was justified in writing: "the idea occurred to me that it was possible to devise an instrument which should do for the eye what the phonograph does for the ear, and that by a combination of the two, all motion and sound could be recorded and reproduced simultaneously" (Clark, 1977, p. 171). Edison's kinetoscope was accordingly also called the optical phonograph. This fact is not only significant for Edison's practical kinetoscope, but it is also theoretically significant: as with Demeny, the first experiments in the direction of multimedia were already happening at the end of the nineteenth century. After the individual sensory channels had been physiologically measured and technically replaced, what followed was the systematic creation of multimedia systems, which all media have since become. What emerged were simulations or virtual realities, as they are now called, which reach as many sensory channels as possible at the same time.

Edison built the first film studio in the history of media, his so-called Black Mary, precisely for this purpose. This Black Maria was, in memory of the *camera obscura*, a large box, which could be turned in the direction of the sun for the purpose of lighting, which was equipped with light bulbs for the same reason, and which had black interior walls so that the illuminated and recorded actors – the first in the history of film who actually performed short, fictional scenes – could act in front of a uniform background. In contrast to the arts, media always play against the backdrop of noise, which in the case of Edison's optics was a black painted wall. The acoustics were more of a problem, as Edison wanted to record the picture and sound at the same time. Without microphones he had difficulty in bridging the distance between the actors' mouths and the phonograph trumpets without disruptive acoustic background noise. The synchronization of "movies," as Edison already called them, with the phonograph cylinder also created problems during playback. Apparently, it was historically still too early for the audiovisual *Gesamtmedienwerk*. Edison also confirmed this after several kinetoscope experiments, when he told students that instruction through film and vision (and not through gramophony) would soon replace instruction through books.

Film, in other words, began at least technically as silent film, and it did not combine all three of Edison's innovations – film, light bulb, and phonograph. It is probably a historical rule of post-print media technologies that individual and isolated sensory channels must first

163

be completely and thoroughly tested before any thought about connecting them is at all possible. The multimedia system of kinetoscope and light bulb was thus the only one that took hold – first in Berlin, with a rather inconsequential film presentation by the Skladanowsky brothers on November 1, 1895, and soon thereafter, namely on December 28, 1895, in the Indian Salon of the Grand Café on the Boulevard des Capucines in Paris, where the brothers Auguste and Louis Lumière gave their first public film demonstration before a paying audience with worldwide results.

The two Lumières – whose surname has already been commented upon thousands of times – really brought light to film. This was for the simple reason that they upgraded the equipment, which had been purchased from Edison along with Edison's film standard, in one small way: they projected films onto a large screen for a paying mass audience, who gathered around a single vision like in the old theater. Above all, however, they developed along with what has since been called the *cinématographe*, or cinema for short, a device that can record, copy, and play back moving images. The cinematograph recorded films when it worked with a lens like a *camera obscura*, it copied films when the lens was replaced with simple sunlight, and finally it projected films when sunlight was replaced with a light bulb behind the film roll. Every spectator paid one franc, and in exchange they simultaneously saw exactly what the other spectators saw. With the phonograph, such distribution was more or less natural due to the fact that the ear cannot be closed, but with film it had to be constructed. The Lumières typically employed front projection with a strong lamp illuminating the celluloid from behind, which has since become standard practice. Once, at the *Exposition Internationale* in Paris, they successfully experimented with front and back projection at the same time. In an enormous hall with spectators sitting everywhere, they were all supposedly able to see the film well. For this reason, a screen was first submerged in water before every showing and stretched across the middle of the room – then half of the audience was able to watch the film projected from the front and the other half from the back due to the water.

If this switch between front and back projection reminds you of someone, so much the better, for the logical coherence of film history then becomes clear. Daguerre's diorama of Vesuvius had switched between day and night views by switching between front and back projection in precisely the same way. Already for this reason it could be no coincidence that film, despite Edison, did not originate in the USA. The conversion of the representative arts into optical media

164

took place in a country that had long known the arts as such. I mean, to quote *The Song of Roland*, *"la dulce France."* Like Daguerre, the father of the Lumière brothers was also originally a painter who became a photographer. But it was precisely because of this technization of his prior handwork that he was concerned that the technology could do without him as a professional photographer simply because people would go on to photograph themselves. Lumière thus directed himself and his sons away from photography and towards the manufacturing of photographic materials . . .

The history of the development of this medium in one generation was continued by his sons, who proceeded no longer simply to take photographs, but also to supply their father and the business in general with better photographic negatives (Télérama). They were therefore both scientists and industrialists who developed a method of storing and projecting moving and thus living people, as well as the first technique of making corpses imperishable and thus storable using formaldehyde. There is no better way to illustrate the connection between media technology and physiology that existed in the nineteenth century.

The contents of the films that the paying spectators on the Boulevard des Capucines saw also resulted from the Lumières' occupation. The first film to be privately screened, which to my knowledge is now lost, was shown at an annual meeting of the French society for photography, where President Marey and all the other scientists were able to watch themselves (Zglinicki, 1979, p. 171). The first film to be publicly shown, on the other hand, showed the employee side of this science, as it presented the Lumières' workers streaming out of the factory gate in Lyon during a shift change. It is characteristic for the difference between media and arts that this film did not present an infantile or humorous but still planned and composed American plot, as with Edison, but rather it was taken purely from everyday life. The Lumières had no Black Mary to bring fictions into the world, but rather at the beginning they only made daylight recordings, which made them the founders of documentary film.

Another confrontation, which was more in keeping with the Grand Café's Indian Salon (the name implies that it was designed for exotic wonders), was experienced by the 35 spectators at the public première. Among the Lumières' documentary films was *L'arrivée d'un train à la Ciotat*, or the arrival of a train at the station of a French city on the Mediterranean, which has since become famous. The favorite toy of the nineteenth century thus entered, the old Renaissance perspective went into effect as usual, and the locomotive on the screen became larger and less well defined until the spectators reportedly fled the Parisian café in fear. Without planning it, the Lumières had actually transformed the spectators not into targets of their fixed camera, but rather into (as Virilio formulates it) targets of the imaginary locomotive. When the American director Griffith later proceeded to put the film camera itself into apparent motion and directly approach the actors with it, this shock effect supposedly increased: the spectators could allegedly only explain the enormous close-ups of faces that filled the screen by concluding that Griffith had literally decapitated the actors' heads.

In the eyes of these deceived spectators, and behind the back of silent film producers who did not have such shocks and murders in mind at all, cinema thus transformed from the very beginning into an illusionary medium. In contrast to the scientific experiments of a Muybridge, which were supposed to replace everything imaginary or figurative in the eyes of people with the real, and in contrast to the phonograph as well, which could only reproduce the reality of noise for lack of cutting or editing possibilities, a new imaginary sphere emerged. It was no longer literary, as in the Romantic period, but rather technogenic. Tzvetan Todorov's theory that the fantastic in literature died after it was elucidated by Freud and psychoanalysis (Todorov, 1973, pp. 160–2) is partly false: the fantastic experienced a triumphant resurrection through film.

Nothing could attest to this more perfectly than the fact that a certain Georges Méliès, who had once been the director of the Robert-Houdin theater, was among the many people who purchased a cinematograph from the Lumières. Robert-Houdin, whom Hans Magnus Enzensberger had appropriately evoked in one of his mausoleum poems, was neither a playwright nor a director, but rather the most famous magician and escape artist of the nineteenth century. His grandson consequently transformed magical artworks into modern tourism by inventing the French specialty of *son et lumière*, a Bengal sound and light show for old castles that designates tourism as the worthy heir of absolutist lighting effects. As the heir of Houdin, Méliès consequently transformed the documentary film into the modern fantastic. He invented a vast number of film tricks, but I will only focus on two elementary ones: backwards projection and the stop trick.

Méliès employed backwards projection perhaps most successfully in his film *Charcuterie mécanique* (Mechanical Delicatessen). A pair of scenes were filmed in a butcher's shop, and they recorded in sequence the slaughtering of a pig, its dismemberment, and the

166

production of a finished sausage. These same scenes were shown at the screening, except that within each scene the last frame had been made the first and the first had been made the last. In the spellbound eyes of the spectators, the resulting film showed a finished sausage transforming back into the corpse of a pig and the corpse then transforming back into a living pig. For the first time in history, the resurrection of the flesh – this 2,000-year-old proclamation – actually came to pass in real life. The ability of film to visually produce apparent continuity could not be demonstrated more triumphantly, as his working principle – the cutting up of living movements into lifeless, static frames – was blatantly disclosed in the form of the mechanical butcher shop, yet the process was nevertheless reversed again in the imaginary sphere. It is therefore precisely because film works in physical time, unlike the arts, that it is in a position to manipulate time. According to a wonderful dictum of the physicist Sir Arthur Eddington, the irreversibility of physical time or the constant increase of entropy, which is a result of the second law of thermodynamics, is shown by the impossibility of films like *Charcuterie mécanique*, which reverse the time axis.

The time reversal trick could also be performed with a sound recording on cylinder or record instead of a film, as Edison had already experimented with playing noises or voices backwards. However, no sound storage device prior to the tape recorder would have been able to keep up with the second trick introduced by Méliès. He apparently discovered the so-called stop trick by accident while filming a Parisian street scene with a hearse. He always filmed with a tripod, which represented for him the unchangeable and therefore illusionary position of the spectator. The celluloid roll ran out in the middle of the scene, however, as the length of these rolls was still not sufficient for feature films before the turn of the century. Without moving the camera from the tripod, a new roll was inserted and the filming continued. Upon projecting the finished product, Méliès was astonished to find that the spectator did not notice the temporal disruption at all (which would be entirely out of the question with the abrupt interruption of a recorded noise). The pedestrians and vehicles passing by on the street had been removed as if by magic, and they had been replaced with other pedestrians in other positions on the street. Méliès immediately incorporated this principle or trick into his next film: *L'escamotement d'une dame*, or the vanishing lady, demonstrated that under media-technical conditions a Robert-Houdin is no longer necessary to conjure people and more specifically ladies away from the stage. And if "lady" is interpreted as Mother Nature, as would be appropriate in classical-romantic literature, then film tricks signify simply a female sacrifice, which has since liquidated all of nature. With the stop trick, film incorporated its own working principle, namely placing cuts in sequence, into its narrative. All that remained was to explain Méliès' technical discovery as the focus of a particular profession, and the job of cutter was born.

So much for the origin of silent film, which from the start had already measured out the entire range of possibilities between accidental realism and illusionary theater, and between documentary and feature film. The only element that was still missing in order to plumb all these possibilities was a moving camera. This task was left primarily to American directors like Porter and Griffith. Zglinicki showed insight for once when he noted that the moving camera, with the possibility of tracking in for close-ups and tracking other moving objects, gave birth to the urcinematic genre of the western (Zglinicki, 1979, p. 492). Classical western scenes that depict enemies, primarily Indians on moving horses, from the point of view of a moving wagon, completely dismiss Méliès' fixed theatrical perspective; they sacrifice the constraint of the spectator's gaze, which was necessary for them to be deceived by stop tricks, in exchange for another and more mobile illusion, which Einstein had described not by chance at the same time, namely in 1905, in his special theory of relativity. Einstein's theory begins with the impossibility of determining, when two movements are relative to each another, such as when two trains pass each other, which movement is virtual and which is real.

The mobile illusion called film thus changed thinking and feeling. Stéphane Mallarmé, who traced literature back to its 24 naked letters without any optical or acoustic illusions, was once asked in a survey what he thought of illustrated books. The answer: he did not think much of them, because readers of Mallarmé, unlike readers of Hoffmann, were not supposed to hallucinate, but rather simply to read; whoever needed illustrations should put away their books and go to the cinema instead (Mallarmé, 1945, p. 878). In another interview, the same Mallarmé was asked how art could improve technology, and he answered by suggesting that the driver's seat in the recently invented automobile be relocated to the rear, behind the automobile owners, who at that time were upper class, and the front windshield be enlarged as much as possible. Without perceiving their own movement at all, and without any visual obstacles such as the driver, these automobile owners would then be fully immersed in the illusion that the streets and landscapes in front of them were opening up and increasing in size as they rode past (Mallarmé, 1945, p. 880).

Mallarmé had correspondingly written a prose poem that immortalized this automobile in the poetic or overly poetic disguise of a
self-rowing boat. The aforementioned boat floats down the river
where Mallarmé's own country home in Valvins was located, slips
through white water lilies (from which the poem takes its title), at
first with the practical goal of carrying the rower, Mallarmé, to a
lady friend. The visual spectacle of a gliding perspective delights the
poet so much, however, that he soon abandons his visit and instead
celebrates the lady as the "absent one" – Mallarmé's most solemn
poetical category (Mallarmé, 1945, pp. 283–6). Mallarmé's poem
thus does exactly what Meissonier (presumably his neighbor) did
with his mobile railroad carriage, which proves that 15 years before
the first tracking shots were made the dream of them already existed.
The mobile images of film are inextricably linked with the new
automobiles and the only slightly older railroad journey. Further evidence, namely from instantaneous photographer Ernst Mach and the
originator of psychoanalysis Sigmund Freud, will follow later today.

The systematics of optical mobilization is in the first instance more
important. First, the step from a static to a mobile camera had eliminated every similarity between film and the ancient art of theater.
From Plato's cave to the peep show theater, spectators were and are
fixed in place – not out of old European sadism, but rather because
before the invention of computers the calculation of moving gazes
would have exceeded all computational capacities. Macroscopically,
film does not actually alter the fixed position of the spectators' bodies.
Microscopically, on the other hand, an apparatus representing their
own eyes performs random cuts and movements, from which the
spectators cannot distance themselves as they have no other possibilities of optical control in the darkness of the cinema. For this reason,
according to Edgar Morin's brilliant formulation, *"the spectator
reacts before the screen as before an external retina telelinked to his
brain"* (Morin, 2005, pp. 134–5; italics in original).

Before continuing with the actual history of film, I would like
to follow another thread through to the end. The cinema that the
Lumière brothers created on the Boulevard des Capucines in Paris
was not simply an apparatus composed of the cinematograph and a
projection wall, but rather it was one of the many child kidnappers
of Hamelin. The question remains what raised the cinema as pied
piper above the old desires of theater. This question leads us back to
lighting technology.

The peep show theater, this baroque translation of the *lanterna
magica*, was illuminated, heated, and filled with smoke thanks to

wax candles, and the duration of its pieces were thus very limited. The wish for a brighter and cleaner light source arose as soon as chemists first began manufacturing flammable gases and Robertson discovered the carbon arc lamp for special effects. The introduction of gaslight in nineteenth-century theaters not only had the effect of increasing the number of theater fires and deaths to an historically unheard of degree, until the introduction of the iron curtain, which was supposed to protect against such disasters, but it also posed a theoretical problem. It was no longer necessary to be stingy with light, and the stage could be made as bright as desired. The question was thus whether there should still be chandeliers in the auditorium, which was a centuries-old tradition, even though they were no longer necessary for people to see the drama or the opera. The answer given by the well-known architect Garnier as he was building the new Grand Opera House in Paris is noteworthy: according to Garnier, a dimming of the auditorium would be possible, as it already existed in a few Italian opera houses, but it was not feasible. First, opera visitors had to be able to read along during the dazzlingly incomprehensible songs in the libretto of the current opera in order to understand at least some of the plot. Second, as a social event people go to the theater not only to see but also to be seen. (Princes, above all, were always illuminated in their boxes, because for them everything depended on courtly representation or glamor rather than bourgeois illusion.) Third, Garnier argued that it is crucial for actors and the artistic quality of their performance that they see all of the audience's reactions; they thus perform in an optical feedback loop. Fourth, a darkened auditorium would also have the disadvantage that it would not be controllable down to the last corner. Opera visitors who no longer read along in the libretto during a love aria might resort to quite different thoughts or actions.

The morality of the gaslight theater thus immediately bears comparison to the new morality or rather immorality of the cinema. While the new technical light sources were not permitted to change anything in the auditorium at first, there were still experiments with new stage effects. I want to cite two particularly magnificent examples to provide evidence of the training of a cinematic gaze.

In the nineteenth century, England had a very practical mathematician. In 1830, Charles Babbage constructed a forerunner of the first computer, which he actually did not complete but which he employed as leverage to bring British precision mechanics to the same industrial standards as Colonel Colt. In 1846, Babbage sat in the German Opera House in London next to a lady wearing a hat.

The gaslight over the stage and auditorium went on, and the lady's hat and program changed color with a slight tinge of pink. This technogenic accident gave Babbage a revolutionary idea for a revolutionary theater lighting system. He got together with none other than Michael Faraday, the discoverer of the stroboscope effect that had made film possible, to create together a simple new ballet. Babbage and Faraday illuminated the stage with dazzlingly brilliant limelight lamps, which an English lieutenant had invented for the purpose of sending military signals, and in front of each lamp they built rotating glass filters in various different colors. Babbage then wrote a ballet about natural science research and the rainbow, put all the dancers in white tricots for the duration of the piece, darkened the auditorium, and illuminated the white tricots with changing colors according to each phase in the ballet's narrative. You can and must imagine the result: the dancers' tricots could alternately shine in all the colors of the rainbow without a costume change. Long before Edison's Black Mary – his film studio illuminated by light bulbs – it was the first theatrical piece developed on the basis of spotlights. And unfortunately, Babbage's piece can only be imagined because the impressario dropped the ballet from the schedule shortly before its première for fear of a fire and an audience in flames.

For this reason, the history of spotlight effects, which introduced virtual and thus proto-cinematic movement into the theater, continued in Germany. A dramatist and opera composer lived there who, for anarchistic reasons, loved the thought of a burning theater and audience. I am speaking of Richard Wagner, and unfortunately I will have to spare you countless details. In short, Wagner's newly founded opera house in Bayreuth truly achieved the transition from traditional art to media technology. I will merely point to the many stage directions in the *Ring* that provide for smooth cinematic scene changes and for the burning of the entire theatrical fortress of the gods, Valhalla, at the end of the entire tetralogy, which recalls Babbage's ballet. Another similarity to Babbage's piece is the rainbow that Wagner's gods conjure on the ceiling of the stage at the end of *The Rhine Gold* after a metaphysical fog had previously submerged them in the most terribly deadly grey. Nothing else signifies *The Twilight of the Gods*. In addition to the coloring of scenes through spotlights, there was also a kind of virtual automobility in Wagner's opera even at its first première in 1876: in the libretto, only the new Valkyries ride on old-fashioned, proto-Germanic horses; in technical positivity, on the other hand, a moving *lanterna magica* projects automobile phantoms of Valkyries onto the rear backdrop. The same

171

happens at the end of the second act of *The Valkyrie*, where Wotan and Brünnhilde appear as shadows on the cyclorama over the two fighters, Siegmund and Hunding (Ranke, 1982, p. 47 – this scene is preserved in an image in the *Illustrirte Zeitung* of 1876). In the hallucination scenes, the controllability of theater light, which was first made possible through gaslight, escalates to *lanterna magica* effects (Ranke, 1982, p. 42).

And because Wagner pushed both the acoustic and the optical effects of theater equally to the extreme, the opera house that he built on a green hill overlooking Bayreuth, which looks like an enormous factory for producing arts and crafts, represented a decisive step beyond the nineteenth century in general and Garnier's Grand Opera House in Paris in particular. In the words of a contemporary eyewitness (or rather earwitness) of the first *Ring* première:

> In Bayreuth the darkened room became the objective. It was also an entirely surprising stylistic device at that time. "A completely dark night was made in the house, so that it was impossible to recognize one's neighbors, and the wonderful orchestra began in the depths." (Wieszner, 1951, p. 115)

If media technology must first isolate and incorporate individual sensory channels and then connect them together to form multimedia systems, then Wagner's Bayreuth opera was the first historical realization of this principle. To highlight the music of *The Rhine Gold* overture, there was nothing and no one to see either in the auditorium or on stage. Not to mention the fact that the opera musicians were not visible at all, as Wagner had sunk them in an invisible orchestra pit following the unique model of the theater built by von Claude-Nicolas Ledoux in Besançon around 1770. Even when the curtain was raised over the huge gas-lit stage, the lights in the auditorium remained off – a practice that still continues today. This no longer shocks contemporary cinemagoers, who only purchase Bayreuth tickets in exceptional cases. However, it was a scandal in 1876, as a newly crowned German Kaiser was among the audience at the first première. Wilhelm I thus had the privilege of being the first prince whose absolutist self-representation was confined to his box by Wagner's anarchism. The age of democracy, or rather the age of prevailing illiteracy, had begun. For spectators who wanted to continue reading the opera text, as Garnier suggested, Wagner had to distribute a flyer prior to the first première to warn them that they should already read through the text, as it would be too late and too dark during the performance. *The Ring of the Nibelung* thus ran as a total film, which was not once interrupted by superimposed intertitles as in old silent films. It was therefore no wonder that Wagner's notorious son-in-law Houston Steward Chamberlain suggested that symphonies should also be performed in the complete darkness of Wagner's opera house, meaning that even the actors who were not yet entirely invisible in Wagner's musical dramas should be cut out and instead only large-format *lanterna magica* effects should be projected, as had been done with the valkyries. Through the elimination of all empirical human bodies, a multimedia show would thus emerge.

Hopefully, this overselling of the theater, which Babbage and Wagner had already achieved in the old peep show theater, sheds more light on Edison's light bulb and the Lumières' film projection. It is clear that electric light made the difference between bright and dark consistently controllable, and therefore in this case it could be set absolutely; it is also clear that it had minimized the danger of theater fires or later also cinema fires. From then on, there were only explosions and catastrophes (provided that we are allowed to ignore the ending of *Gravity's Rainbow*) when celluloid films – whose chemistry is for good reason narrowly related to explosives – caught fire under the heat of projector light bulbs. However, the projection of electric light through an otherwise darkened room seems to be the most important thing. It not only established the aesthetics and social pathology of cinema, but it also created a new militaristic way of perceiving the world. It was not until the invention of klieg lights around 1900 that film studios, following the model of Edison's Black Mary, could finally bid farewell to daylight recording. After the rise of film, the theater could also shift its lighting to virtual effects and klieg lights, as Max Reinhardt did in Berlin. The fact that actors today almost always stand in spotlights is therefore an imitation of cinema. And finally, light projection could also modernize warfare. In the Russo-Japanese War, for example, the Tsar's army employed spotlights for the first time to protect Port Arthur, Russia's last eastern Siberian fortress. When the Japanese attacked at night, these spotlights transformed battlefields into lethal film studios (Virilio, 1989, p. 68). If bedazzlement, as I have already said, was a privilege of princes and the powerful during absolutism, enabling them to humble their subjects, it has since 1904 become an actively armed eye that no longer simply optimizes its own perception, like the telescope and the microscope, but also reduces the perception of the enemy to zero.

We will return later to the nexus between war and cinema. For the historical moment of silent film, however, it is sufficient to study the aesthetics of this actively armed eye and the sociology of its subjects. As we are pressed for time, I will skip over countless opinions on silent film that range from Georg Lukács to Béla Balázs and limit myself instead to a very early silent film theory: Hugo Münsterberg's slender book *The Photoplay*, which was published in New York in 1916 and which connects most elegantly to my questions about cinema, theater, and lighting. Our last session ended with Edgar Morin's aphorism that the cinemagoer catches sight of his own immeasurably magnified retina on the projection screen. Münsterberg had proved precisely that already in 1916.

Hugo Münsterberg, to introduce him briefly, was a lecturer in experimental psychology in Freiburg im Breisgau and thus a colleague of all those like Fechner, Helmholtz, and Marey, who were present at the birth of film. William James, the donnish brother of novelist Henry James, became acquainted with the young lecturer at a psychology congress in Leipzig, which at that time was still a leader in science thanks to Wundt. Out of pure enthusiasm, James offered him the directorship of the newly founded experimental psychology laboratory at Harvard. At the turn of the century, in other words, universities followed the solitary example of Edison, the self-made man. In his laboratory, for example, Münsterberg had already taught the young student Gertrude Stein experimentally about automatic or surrealistic writing 20 years before she achieved literary fame. But writing was only one of countless cultural techniques that he measured according to all manner of psychological and physiological parameters, with the actual goal of optimizing all of these techniques ergonomically. The upcoming assembly-line work, as it was implemented by Henry Ford during World War I, required that every bodily movement, even inconspicuous movements like writing, proceed optimally. I would add here that Gilbreth, an American colleague of Münsterberg's, conducted such ergonomic measurements with the help of slow-motion films. The ex-Freiburger was not as practical. He only went to film studios, which at that time were located in New York rather than under Hollywood's sun, and were as an exception opened to him because of his fame in America. The book *The Photoplay* was Münsterberg's last success, however, because one year later the German Empire declared war on the United States. Münsterberg had tried to prevent this war through fireside chats with President Wilson, but his efforts were in vain. In 1918, he mourned the fact that no president, professor, or American would greet him any longer, and he died of a heart attack while presenting a lecture later that year.

His film book provides an example of applied psychotechnics, as Münsterberg dubbed his new science. In order to explain a modern media technique, Münsterberg cleverly employed an older counter-example: the peep show theater. The result of the aesthetic comparison is that the theater can actually awaken many illusions among spectators, but none of them are physiological. The stage disrupts neither the physical place nor the physical time of the events; in fact, it must simply accept them. (This is the reason why the three unities became a theoretical problem for theater.) On the other hand, the photoplay or feature film is a true psychotechnique, which attacks and modifies the unconscious psychological states of the cinemagoer using strictly technical means.

The examples for this thesis come from the realm of film tricks and montage techniques, which had already been achieved at that time. As if to prove the objective and not just the etymological connection between the notion of montage in film and the notion of montage on the assembly line, the first film theory deals with the effects that close-ups, flashbacks, flashforwards, and reverse shots have on spectators. In the unconscious and therefore uncontrollable act of spectating, psychic acts correspond to all of these effects. The most obvious is Münsterberg's example of the close-up: the protagonist of a film wants to shoot someone, so he takes hold of Colonel Colt's revolver. However, the film camera – the historical descendant of this same revolver – is not satisfied with simply carrying on looking at the hero, which would be the only available possibility in the theater. In fact, as an actively armed eye it tracks in on the hero until only the hand and the revolver fill the entire image in the lens. According to Münsterberg, all unconscious attention functions in the same way: it filters out completely irrelevant image components without noticing. In a similar way, the flashback realizes or implements unconscious memory (which Proust was investigating in literature at exactly the same time), the flashforward realizes or implements unconscious fantasies of the future, and the montage of temporally or spatially separated scenes realizes or implements the functioning of unconscious association in general. All the irresistible and uncontrollable shock effects unleashed by Lumière, Méliès, and Griffith are thus explained using psychotechnics.

Münsterberg's theory thus completes, at least for us, a scientific-historical circuit. As I have repeatedly emphasized, media technologies emerged in the nineteenth century from psychological and physiological research on a very empirical and no longer transcendental human subject. When Münsterberg was writing in 1916, on the other hand, these media technologies were perfect and could, in turn, provide models for psychology and physiology. If unconscious attention is nothing more than a film trick, then humans can be built and optimized instead of being further idolized idealistically. Pynchon's cited film director was indeed mistaken when he provided the comforting assurance that we do not yet entirely live in film. In technical actuality, the scientific experimental film above all changed the realities of life itself. People working on an assembly line perform movements taught them by a film.

Münsterberg's admirable theory also completes a literary-historical circuit. As you may remember, romantic literature demanded – to quote Novalis directly – a "right reader" who develops a real, visual, and also audible inner world based on the words on the page. Novalis' wondrous novel *Heinrich von Ofterdingen* depicts this process, as someone tells the hero about a wondrous blue flower, which the hero never gets to see throughout the entire novel. Under conditions of silent reading, however, the hero lapses into a dream upon hearing the tale, and this dream places the blue flower in front of an inner or hallucinatory eye. At the end of the dream, the flower's petals become clothing and the center of the flower becomes a woman's face. The hero cannot help but remain faithful to that hallucinatory beloved even after her death, for the duration of a romantic's life or at least for the duration of the novel.

Münsterberg responds to this scientifically, and that means scathingly:

> Rich artistic effects have been secured, and while on the stage every fairy play is clumsy and hardly able to create an illusion, in the film we really see the man transformed into a beast and the flower into a girl. [. . .] Every dream becomes real. (Münsterberg, 1970, p. 15)

If film, according to Münsterberg's clear words, simply surpasses literature and theater, then this new media aesthetic has consequences that people other than myself would call sociological. The fact that theater and novel publishing have experienced a crisis in popularity and profitability as a result of the rise of film is still harmless compared to another threat. If film tricks can actually appear to make flowers out of women rather than merely in the reader's imagination, the ideal beloved of all romantic authors and readers dies out. Novalis' Mathilde and Hoffmann's Aurelia, who were only

176

seducible as readers, disappear simply because they exist in the necessary blurring of descriptions and thus as a single indistinguishable ideal figure. They were replaced, to put it simply and dramatically, with empirical-statistical women.

The film star – who, as Georg Lukács already recognized in 1913, actually has no soul or doesn't need one, but who is simply an unmistakable human anatomy – is naturally first and foremost an empirical woman. The *Kinogirl*, as starlets were called in the leading German film magazine in 1911, "constitutes a morally delightful companion to theater actresses, as they have vegetated in Europe since 1650 without receiving Christian funerals during the first centuries of their existence." The magazine *Kinematograph* takes an opposing view:

> How a young woman comes to the theater is clear to everyone interested in either the theater or the young woman. People seldom worry about how a young woman holds her ground on the stage. But the public should know how many sad dramas of life the photograph *hinders* by *showing* dramas from the life of the public. Then the old accusation that the cinema restricts the life blood of the stage would no longer be heard for a long time, and those who perish on the stage would be offered a rescue. (quoted in Schlüpmann, 1990, p. 19)

This grandiose argument is implicitly between women's bodies and the illustrations of women's bodies. Film stars are actually just as erotic as theater actresses, but they merely have to prostitute their images rather than their bodies. For media-technical reasons, therefore, chastity or sanctity dwells within them, which in Hoffmann's *The Devil's Elixirs* was only true of the painted Rosalia and not of her fleshly double Aurelia. In other words, the cinematic pin-up girl deflects palpability just like three-dimensional technologies in two-dimensional space. For this very reason, however, the borders of palpability are no longer drawn. Everyone knows the famous story of how Howard Hughes, the multi-millionaire constructor of military aircraft, also constructed a special bra for Jane Russell's unmistakable anatomy for the purpose of making films.

What is more forgettable and more important, however, is that empirical-statistical women were also cinemagoers. At precisely the same time as Germany's universities first accepted women, and thus the Faustian ideal love, Gretchen, was historically set aside, early silent film cinema was considered to be women's entertainment. This is powerfully shown in Heide Schlüpmann's habilitation treatise, which examines spectator statistics as well as the content of films in Germany prior to 1913. I will only add that according to quite a number of historical witnesses it appears that secretaries in particular went to the cinema. Women's emancipation had not only invented the student as a new profession for women, but also the typist. After the media-technical collapse of literary illusions, all that remained for these typists was the dry and – for male writers in general – too trivial task of working ten hours a day with bare, meaningless letters. Such secretaries escaped the daily grind of their office jobs, as they themselves testified in magazine surveys in the 1920s, simply by seeing a boyfriend and/or going to the cinema every evening, where they were guaranteed not to be threatened by any texts other than the intertitles.

If anyone wants to have printed evidence of the historical primacy of women cinemagoers, they should read Jean-Paul Sartre's autobiography *The Words*. In this book, the old philosopher recalls the equally child-like and literary con-artist that he was as the grandson of an all-powerful writer. Of course, his grandfather, like all of his bourgeois friends, attended the Parisian theater as often as possible in order, as Sartre so beautifully formulated, to be "insidiously prepared for ceremonious destinies" (Sartre, 1967, p. 75). It would have been easy for Sartre to become a Stalinist revolutionary, however, if his young mother had not dragged him into her own passion for film. Women, children, and welfare cases – not only among the Sartres – were thus the principle audience for films until about 1910, when men developed such a fear of this literary desertion that they either introduced film censorship or they invented the masculine auteur film, which essentially amounts to the same thing.

Film censorship was demanded by diverse moral organizations and was ultimately anchored in the law of an empire, although its reigning technophile was naturally also a film fan. There were two different reasons for its existence: the content of films and the social space of the cinema itself. The social space of the cinema realized all of Garnier's fears and all of Wagner's hopes. A cinema in Mannheim, which was the subject of a dissertation written by an early female student of none other than Max Weber in 1913, advertised with the slogan: "Come in, our cinema is the darkest in the entire city!" My task is not to present a lecture on the social history of petting, but I would like to direct your attention to Gottfried Benn's enthusiasm for this darkness (in the early novella *The Journey*) and at the same time articulate a warning: early cinema can certainly be schematized as a feedback loop between erotic film content and the erotic practices of cinemagoers in the same way that romantic literature served as such a feedback loop, but we should always question, as

178

Pynchon does, what authority programmed such loops. As Zglinicki reportedly heard from the mouth of a direct participant, the most famous naturist film, which was made by the UFA (Universal Film AG) after World War I, was intentionally supposed to have only an indirectly eroticizing function; its main purpose was rather to show healthy souls in healthy bodies immediately after a lost war and thus contribute "to military fitness" (Zglinicki, 1979, p. 576). The armed forces had assisted in making the film.

But now I am getting ahead of myself. To understand how a UFA could come into existence at all, we will return once more to the individual stages by which film fell again under state control.

The first stage, as I have said, was the auteur film, which did not emerge in Germany until 1913 at a time when the country was rather dependent on film imports. In France, the *comédie française* had already pounced earlier. After a few years of wild polemics, which Anton Kaes has replicated in his volume on cinema debates (Kaes, 1978), theater people and novelists decided to make peace with the new competitive medium just before the start of World War I. In 1912, the largest German film company and the writers' union signed a joint agreement concerning royalties and copyrights (Schlüpmann, 1990, p. 247). One year later, in 1913, the famous actor Albert Bassermann, who had previously refused every photographic portrait out of the same fear of the camera as Balzac, was persuaded to perform in the first German auteur film, Paul Lindau's *The Other*, in which he played his own double. This culturalization of cinema, in turn, was supposed to abolish everything that films had inherited from fairs and magic shows, as well as some aspects of the social space of the cinema that it had inherited from Wagner's opera and the *chambre séparée*. Following the models of New York, Paris, and London, Berlin also began to develop film palaces, which were hybrid architectures that combined features of both the cinema and the theater. To provide appropriate film content for these palaces, in which both the educated middle-class and the cultural set were able to set foot, novelists had to write screenplays, and theater actors had to be filmed. The two most famous cases were the novelist Hanns Heinz Ewers and the actor Paul Wegener, who made the second German auteur film, *The Student of Prague*, in 1913.

The work of the screenplay consisted first of all in turning literary hallucinations into cinematic positivities. With his typewritten screenplay – a text that was thus reduced to naked letter sequences – Ewers did this brilliantly, while Gerhard Hauptmann reportedly failed at the same task. Second, it naturally involved replacing all of

179

the fallen or servant women of early women's films with a man – preferably an educated one. The sciences that had historically made film possible were thus incorporated back into it again. For example, Ewers' student suffers from a hallucination of a *doppelgänger* that, as early psychoanalysis immediately recognized, can only be explained through psychoanalysis. Other auteur films, like *The Other* or *The Cabinet of Dr. Caligari*, feature sciences like criminology, which since Bertillon focused on securing photographic evidence, and psychiatry, which had already in Marey's time produced the first serial photographic snapshots of patients. Münsterberg would have had no problem in finding his own experimental psychological premises realized or even implemented in *The Student of Prague*.

Beyond the reconciliation between literature, science, and film, however, the auteur film also had to mend the rupture between film and theater. The famous theater actor Paul Wegener thus gave his first experiment in front of the camera the lovely title *The Seduced*. His cameraman Seeber, the son of a Chemnitz photographer, experimented systematically with Georges Méliès' stop tricks and all sorts of double exposures to create what he called "absolute film":

> Naturally [Seeber wrote in 1925 in absolute unison with Münsterberg] an entire film will never be absolute, but certain scenes within a large film that depict an internal procedure – a legendary, fairytale-like or fantastic procedure – can be produced on the way to absolute film. Such a film . . . demands a complete conversion of the screenwriter – for once the word "poet" can really be used here – and actually a poet who also understands how to translate his fantasies into technology. He must be able to conceive of the different parts of an absolute film image. He must not only specify the procedures objectively, but they must also be fixed temporally. The screenwriter of the future – and I am firmly convinced that absolute film has a future – will have to write like a musician writes his score. And just as a musician orchestrates his acoustic creation, the film author will also have to write a kind of technical score that enables the photographer to follow his fantasy. (Seeber, 1925, p. 95)

In *The Student of Prague*, Seeber's "absolute film" consequently amounted to the presentation of film as film. All of the double exposures and stop tricks that Seeber had learned from Méliès, which he augmented through his passion for American klieg lights, only served the goal of confronting the theater actor Wegener with himself as an "other" or double. This "other" looked completely the same, but he was missing any inwardness or facial expression. In this way, he

180

seemed like the idiotic and that means cinematic negative of the positive theater star. In other words, the *doppelgänger* trick represented a
film of making a film. A famous actor died simply because there was
a copy of him on the screen. Remember why Garnier had refused to
dim the lights in the auditorium: with invisible spectators the actors
could no longer exchange any optical or gestural signs of approval or
understanding. But this interruption of all feedback loops between a
body and its doubles – whether in the mirror, in one's own internally
stored body image, or in the approving eye of the other – precisely
defines technical media. You do not recognize tape recordings of your
own voice because only the acoustics of the exterior space remain,
while the feedback loop between the larynx, Eustachian tube, and
inner ear does not work in front of the microphone. The number
of early horrified witnesses appropriately shows that people did not
recognize their own moving doubles. Maltitz' comedy *Photography
and Revenge* had already demonstrated how the camera replaces
beautified portraits with the faces of criminals; cinema pushed this
alienation effect even further. The protagonists of novels by Vladimir Nabokov and Arnolt Bronnen, who had become film extras or
even stars, experienced the shock of seeing themselves on screen in
the cinema. For men like Freud, who neither went to the cinema nor
read about it in his books, the same experience could happen in a
train compartment. As the mirrored door of a first-class bathroom,
which at that time was still reserved for the upper class, suddenly
moved, Prof. Sigmund Freud saw according to his own confession
"an elderly genteman" whose appearance he "thoroughly disliked"
yet only later painfully recognized as his own mirror image (Freud,
1953–74, XVII, p. 248).

Beyond all examples of historical scientific anecdotes, this fear of
the double functioned as the social Darwinist principle of selection.
To begin with, the actors who survived it became film actors, while
the others dwindled away together with their medium until they eventually became the subsidized elite they are today. Second, Stevenson's
novella about *doppelgänger*, *Dr. Jekyll and Mr. Hyde*, became one of
the most frequently adapted stories of all time. Third, media-technical
selection principles never remain limited to the art establishment.
The conditioning of new technogenic perceptual worlds not only
concerns producers, but also consumers. Michael Herr, the drugged
war correspondent, reports that during the Vietnam War there were
entire companies of an elite American unit, the marines, that were
only prepared to go into battle on the rice fields, and that means to
go to their deaths, when one of the countless television teams from

ABC, NBC, or CBS were already there waiting and ready for action (Herr, 1978).

World War I had already invented this beautiful death of a double, which the evening news would then celebrate before the eyes of astonished parents. This was the second phase of the domestication of film. For literary scholars, I can only point out in passing that whenever Lieutenant Ernst Jünger describes an encounter with the enemy in his war journals and novels, which was extremely rare in the trenches, he names this enemy his own double. The historical reason for such hallucinations is even more significant: namely, Jünger was only one of millions of other trench warfare soldiers that World War I made into the first masculine mass film audience (and into the first mass radio audience). The phase of women's films was thus in the past. In the communications zone located directly behind the three trench systems, entertainment films were shown for all of the armies during World War I, which also led to film stars like Henny Porten moving into dugouts as pin-ups (Virilio, 1989, pp. 25–6). Without film recordings, the sensory deprivation of soldiers, who were only permitted to see tiny sections of the sky over their trenches for four years (if they survived that long), would have resulted in very cinematic psychoses. It was only through the artificial storage and input of moving pictures that armies of millions were supplied with morale boosters.

Behind the new eroticism (the so-called male fantasies, to quote Klaus Theweleit) there was thus a new war technology. To begin with, it soon became clear to all of the participating nations that world wars could no longer be won without the support of world opinion, or at least published opinion. This publicity dimension of world war strategy benefited the allies, because Great Britain, France, Italy (after 1916) and the leading film-makers, the USA (after 1917), all belonged to the opponents of the so-called central powers. Films presenting a world-destroying and virgin-defiling image of Kaiser Wilhelm II were thus exported to all the neutral countries. These films did not really bother the Supreme Army Commanders of the German Empire, who were still very Prussian and proper, but it did worry the new, technologically savvy team of Hindenburg and Ludendorff, who were ordered to turn the deadlocked war around in 1916. As the strategic head of the Third Supreme Command, Ludendorff thankfully took the advice of a Leipzig industrialist, who in the interest of the "Made in Germany" brand and his own *Illustrirte Zeitung* had already been demanding worldwide film propaganda for the emperor and the empire for years (Zglinicki, 1979, p. 389). Unfortunately, it

182

still remains uncertain whether Ludendorff also listened to Ewers, the successful novelist and screenwriter who was traveling in the USA during World War I and was eventually interned. In a long and still unpublished typescript, Ewers had criticized the idea of German foreign propaganda as pure idiocy. In 1917, however, things suddenly changed: an office for images and films (*Bild-und-Film-Amt*) was founded within the Supreme Army Command – the most sacred Prussian military tradition since Scharnhorst and Gneisenau – which was given the name BUFA due to the fashion for abbreviations during World War I, which has since become the norm. In its 900 cinemas on the front line, BUFA commandeered all the films, projectors, and projectionists that had delighted Lieutenant Jünger in his Belgian base.

But that was not enough. In his capacity as Quartermaster General, Ludendorff wrote an official letter on July 4, 1917 to change BUFA to UFA through the omission of a single, unimportant letter (unimportant since the advent of film). In his general staff-like clarity, Ludendorff sent the following plea "to the Imperial Ministry of War in Berlin":

The war has demonstrated the overwhelming power of the image and of film as instruments of enlightenment and propaganda. Unfortunately, our enemies have exploited their advances in this area so fundamentally that we have suffered heavy losses. For the further duration of the war, film will not lose its significance as a tool of political and military propaganda. To ensure a happy ending to the war, therefore, it is absolutely necessary for film to have the greatest impact everywhere that German influence is still possible. [. . .] What means are to be employed? Because only the absolute majority is required to influence a corporation, it is not always necessary to purchase all of the shares [of a company]. It must not be known, however, that the state is the buyer. The entire financial transaction must be performed by a competent, influential and reliable bank that is unconditionally loyal to the government. The negotiators should not be permitted to know the true identity of the agent's client. (Zglinicki, 1979, p. 394)

Nothing more came of Ludendorff's strategic goal – a happy ending to the war – even though Jünger's world war novels rave about precisely such a resolution. However, the tactical goal of the Supreme Army Command was achieved. Without the state being recognized as the string puller, two apparently private institutions – the Deutsche Bank and the gramophone company Lindström, where Kafka's eternal fiancée Felice Bauer had risen up the ranks from typist to authorized representative – created a company out of various private

German and Danish film companies, which they named "Universal Film AG" or UFA for short. The military industrial complex thus performed its own metamorphosis: BUFA, a branch of the general staff, became UFA, a large-scale industrial film company that appeared to be privately owned on paper, yet was at the same time always half state-owned. UFA continued to produce war and peace propaganda in Babelsberg just south of Berlin until March 1945, when Marshal Zhukov's last offensive at dawn at the Oder front began to blind the remaining German defenses with anti-aircraft spotlights (a tactic reminiscent of the Russo-Japanese War). But even after the Red Army marched into Babelsberg, UFA only needed to change two more letters in its name to start producing film propaganda for the German Democratic Republic as DEFA. That is how long-lived or lifeless the power of the state is . . .

So much for propaganda or the face of a war that did not end until November 1989. The history of propaganda, which we have traced from a papal institution in 1662 to a military agency in 1917, still does not deal with the real problem of war and cinema. The fact that it was ever necessary to entertain soldiers in their dugouts or influence neutral countries in their indecision is only a negative and therefore indirect way of saying that modern wars are no longer visually reproducible. At the same historical moment when film made the motion of a bullet in flight visible, no matter how fast it was going, the technologies that had made film itself possible in the first place disappeared into strategic invisibility. It was the machine gun, this generalization of Colt's revolver, which imposed this invisibility during World War I. This serial killing machine, which had originally been developed and employed only against reds, blacks, and yellows, now turned on its white inventors. Due to the danger of being immediately shot in the head, soldiers were forced to disappear under camouflage and into trenches, and they no longer saw anything between the fronts except for their piece of sky and possibly hallucinations of a Madonna or pin-up figure. The epoch of silent film thus comprised not only millions of spectators, but also millions of invisible people. A world war that demanded worldwide and thus analphabetic and thus silent film propaganda, had no propaganda material to offer at all. To film its material battles, one would also have to be able to film white noise.

This fact frustrated Griffith, the most famous American director of his time. He landed in Europe, planned one of his crowd scene propaganda films, went to the trenches, and saw that the battlefields were empty (Virilio, 1989, pp. 14–15). So Griffith then built a

184

gigantic studio with gigantic but perfectly filmable trenches and simulated a world war as if national or civil wars were still the order of the day. From the very outset, the fictional battle scenes Griffith recorded from the obsolete panoramic perspective of a field commander laid themselves open to the judgment of Schlieffen, who had already prophesied in 1909 as chief of staff of the imperial German army that there is nothing more to see in contemporary wars: the front had become vast and incalculable, and for security reasons field commanders had already had to exchange their hill for a bunker.

In Germany, the situation was much the same for the Berlin entrepreneur and film director Oskar Messter,[8] an "old master of cinema technology" as Zglinicki so old-masterfully describes him. Messter came to film as the son of a precision mechanic and optician, who had constructed (entirely in our sense of the words) electrical spotlight installations for military parades and theaters. Messter himself began manufacturing the newly invented X-ray equipment for invisible light that was beyond even ultraviolet. After becoming acquainted with the Lumière apparatus, Messter proceeded to set up a German film industry and an artificial light workshop, acquired 70 patents, and founded various firms that were ready to go into production, which all merged into UFA in 1918. While it was normal in the age of silent film to record films at horribly slow speeds and then project them in the cinema at much faster speeds to save time, Messter fought for a standard frame rate of 24 hertz for recording as well as playback, which sound film was then supposed to enforce (Zglinicki, 1979, pp. 256–66).

Oskar Messter's film company became historically significant when World War I broke out, as Messter's various firms were made into subdivisions of BUFA. Their rather monopolistic orders from the government were to film newsreels of the war front and then project them on the home front to boost the morale of workers and the wounded. This order led to similar difficulties as Griffith had experiences. Even though there was nothing to see or film in the trenches, the military prohibited shooting on location simply because they did not want to supply the enemy with free intelligence reconnaissance. Like Griffith, therefore, Messter's newsreels had to simulate battle scenes at the base, which unfortunately lacked the desired propaganda effect. As a military hospital chief wrote to Ludendorff, "watching the German war newsreels has an important medicinal effect on the wounded.

---

[8] In the original German text, Kittler incorrectly refers to Oskar Messter as "Oskar Meester."

These gentlemen tell me that they have never heard such thunderous laughter as when those cinematographic images from the trenches and 'from the front' are shown – but laughter is an important remedy" (Zglinicki, 1979, p. 390).

Unlike Griffith, however, Messter learned a lesson about experimental film from this propaganda film disaster. If the trenches simply could not be perceived optically and were also not permitted to be perceived militarily, the only remaining path for film was the vertical path. With the failure of Schlieffen's plan at the battle of the Marne, French Marshal Joffre led a successful counter-strike in August 1914 on the basis of photographic records made by reconnaissance aircraft. After this battle, when the enemy armies disappeared into their trenches, the need for air reconnaissance became even more pressing. A few photographic and film records from the vertical could disclose invisible soldiers, camouflaged artillery positions, and unnoticed rearward connections to the enemy. For this reason, the reconnaissance pilots of World War I represented the origin of all air forces long before bombers and fighters.

In order to help German reconnaissance pilots, Messter constructed his patented "target practice device for the detection of deviations by means of photographic records" (German Reich Patent Office, Patent Specification No. 309108, Class 72 f, Group 7, July 18, 1916). According to the patent, this device was placed in the exact position of the machine gun in a fighter plane, with the aim of helping to monitor the precision of the machine gunners in real time for each individual shot. As Messter put it so beautifully, he employed "a cinematographic recording apparatus whose running gear was propelled by clockwork and whose visual field carries crosshairs, whereby a targeting set-up like that of the machine gun is to be arranged parallel to the visual axis." As you can see, the structural correspondence between perspective and ballistics became a technological reality by World War I at the latest. Messter's ingenious construction, which photographed at least 7.2 million square kilometers of combat area using millions of kilometers of roll film[9] (Zglinicki, 1979, p. 273), could only be improved by combining shooting and filming, serial death and serial photography, into a single act. This was accomplished by a French reconnaissance pilot who relocated the visual and ballistic axis of both the machine gun and the camera to the axis of the propeller (Virilio, 1989, p. 18).

---

[9] Kittler misquotes Zglinicki when he says that "millions of kilometers" of film were used in World War I. Zglinicki's actual figure was only 950,000 km.

This coupling also took place in Germany. Guido Seeber, who built the Babelsberg studios in 1911 and filmed *The Student of Prague* there in 1913, was drafted two years later in 1915 and sent to the experimental seaplane station in Warnemünde (not to say Peenemünde). There, he established at the same time a central educational film and photography hire service. Their first successful scientific film showed with the help of X-ray photography that countless airplane crashes were caused by lead balls built into the wooden propellers for balance; at high speeds these balls fly out and smash through the airplanes like bullets.

In a positive inversion of this negative test result, so to speak, Seeber also constructed a machine gun sight for fighter planes, which was supposed to optimize the machine gun firing rate. And just like the French reconnaissance pilot, Seeber combined this machine gun sight with a small film camera, which also shot film frames whenever the machine gun fired.

Strategically, therefore, filming and flying coincide. McLuhan sums this up succinctly in *Understanding Media*: "It was the photograph that revealed the secret of bird-flight and enabled man to take off" (McLuhan, 1964, p. 174). There was a reason why Marey had also studied the movements of bird wings and why photographers like Nadar had taken pictures from hot-air balloons and passionately fought against zeppelins, supporting instead bird-like – or "heavier than air," as it was called at that time – plane constructions. Gabriele D'Annunzio, the decadent novelist and fighter pilot, already demonstrated in 1909 to a woman sitting next to him on an airplane that by approaching from the air the cathedrals and castles of Italy could be magnified or reduced by any amount and thus also visually destroyed. This discovery, just before the outbreak of the First World War, which D'Annunzio himself only turned into a novel, also unleashed the tracking shots of the first world-famous period film *Cabiria*, which D'Annunzio contributed to as adviser and allegedly also screenwriter. During the war, his flight squadron "La Serenissima" proceeded to fly from Venice across the Alps (which was quite dangerous at that time) to mount an attack on Vienna. These attacks did not consist of bombs, as D'Annunzio explained to the Viennese on propaganda flyers dropped on the city, but rather the Austrians were permitted to remain alive so that they would be able to overthrow their emperor more effectively.

Historians only later recognized that the Viennese-Venetian air war scenario had already been attempted half a century before D'Annunzio, only in reverse. You will recall that Field Marshal

Lieutenant Baron von Uchatius started projecting stroboscope draw-ings in 1841, and somewhat later he also began manufacturing explo-sive Uchatius powder. In 1849, after a civil revolution that also promised freedom to the Italian and especially the Venetian subjects of Austria-Hungary, an Austrian General based in Mestre besieged the rebellious Venetian republic of Serenissima. To the general's chagrin, it appeared that Venice's lagoon prevented it from being captured or even fired on by the artillery. That is, until Uchatius and his brother, two artillery lieutenants from Vienna, then made the world-historical suggestion of attacking from the air: bombs "were to be carried over the city in hot-air-filled balloons made of paper, which would be made to rise if wind conditions were favorable. Prior to being released from the balloon, the 30 pound bomb was supposed to be set [with a time fuse] according to the strength of the wind at that moment. If everything went according to plan, therefore, the impact of the explosives had to take place roughly where it was expected," because the "range" of the bombs "far exceeded the range of artillery at that time" (Kurzel-Runtscheiner, 1937, p. 48). The wind actually very rarely helped, but between June and July 1849 a few of the 110 Uchatius bombs manufactured in Vienna did indeed explode over the astonished Venetians. A field marshal lieutenant, who invented cinematic projection, thus also had to invent the projection or throw-ing of projectiles.

Once these bombs or later planes were manned, the cinematic high-angle shot was born. So World War I not only produced the new professions of reconnaissance and bomber pilots, but also a new kind of film director. These directors had all previously been fighter pilots, and on the basis of their technologically altered visual perception they also revolutionized the entertainment medium of film. Jean Renoir, the director of *Grand Illusion*, was a fighter pilot, just like Howard Hawks, who filmed his war memories in 1930 as *Dawn Patrol*. The clearest example of this nexus between air combat and cinema, however, was Dziga Vertov, the Soviet director, film theorist and above all Lenin's propaganda conductor. Vertov's so-called "rules" for experimental film, which cancelled out all bourgeois infatuation with images, began first with a "General instruction for all tech-niques: the invisible camera." Eight individual points then followed. No. 1 was: "Filming unawares – an old military rule: gauging speed, attack." No. 6: "Filming at a distance." No. 7: "Filming in motion." And finally No. 8: "Filming from above" (Vertov, 1984, pp. 162–3).

So much for Vertov's combat rules as general staff officer – and now, in a free adaptation of Ernst Jünger, the aerial combat film as inner experience of a Vertov demoted to become his own front-line soldier:

> I am the camera's eye. I am the machine which shows you the world as I alone see it. Starting from today, I am forever free of human immobility. I am in perpetual movement. [. . .] I approach and draw away from things – I crawl under them – I climb on them – I am on the head of a galloping horse – I burst at full speed into a crowd – I run before running soldiers – I throw myself down on my back – I rise up with the aeroplanes – I fall and I fly at one with the bodies falling or rising through the air. (Virilio, 1989, p. 20)

In other words, the experimental and entertainment films made with a camera that was no longer only mobile, like Griffith's, but also truly unleashed through tracking shots, simply converted the perceptual world of World War I into mass entertainment. The same thing also happened incidentally in the new media art form known as the radio play, which European civilian radio developed in 1924. And because the war dead returned as an acoustic barrage in postwar radio plays and optical air combat in postwar films, the large cinema palaces between Hollywood and Berlin were also constructed like giant mausoleums. After the European monarchies fell and the old conspiracy between state and church propaganda disintegrated, these cinema palaces became churches of state propaganda that no longer praised a king by the grace of God, but rather (to adapt Lenin freely) war technology and electrification (Virilio, 1989, p. 28).

But it seems to be a law of media history (at least for Berliners) that new applications are secondary compared to new circuit technology. The lessons that film directors learned from World War I pale before the lessons learned by electrical engineers.

The lesson was that film could stop being silent. The technologies of World War I led to sound film, which leads us to the next chapter of these lectures.

3.2.4 Sound Film

The history of sound film has to begin with the assertion that silent film was never silent. Edison had already designed a link between the kinetoscope and the phonograph, which could have been built at a pinch despite problems with sound recording from a distance and synchronization. Many hobbyists and tinkerers followed Edison's lead and attempted to couple a half-mechanical, half-electrical optics

189

with a purely mechanical acoustics prior to World War I, but without any appreciable results. There were also experiments that attempted to connect silent film and electromagnetic sound recording, namely Poulsen's telegraphone. Although these experiments did not yield any concrete successes, they were theoretically significant because they established the principle of magnetic audiotape, which was ready to go into production during World War II. With the audiotape and the cassette, sound recording acquired for the first time the same material format as film: as a roll that allowed variable time axis manipulation, unlike the phonograph and the gramophone. Not only are time reversals possible, as with Edison, but also stop tricks, cuts, and montages, as with Méliès. The simple manipulable acoustics of audiotape led to rock music, as you know, which could then in turn be coupled with manipulable videotapes, and the video clip was born.

But for us, those are all still musical dreams of the future. In the non-experimental everyday lives of people prior to 1929, silent film was never silent simply because films were never presented without some form of accompaniment. The cheapest form was the film explainer, who was often recruited (as we can gather from contemporary adverts) from among the academic proletariat, and who would explain the plot of the film to the spectators while it was playing. Cinemagoers, who had always already been listeners and readers, needed training in the new semiotics of film, which essentially consisted of cuts and montages and thus empty spaces. Kurt Pinthus' *Das Kinobuch* (The Cinema Book), which was published by the expressionist in 1913 and is full of screenplay proposals by his novelist friends, shows what an intellectual step it was to demand films suitable to the medium – film plots, in other words, that were intelligible based on image sequences alone without any intertitles or film explainers. But this *l'art pour l'art* of silent film had not been commercially successful. More expensive forms of sound accompaniment saw to that: music either from records or living musicians, who often also had the honor of generating theatrical sound effects fitted to the scene in addition to sounds on the piano. As a synthesis of two contradictory elements, Greek atmosphere and media-technical noise, Richard Wagner in particular triumphed in the cinema. Wagner not only invented the darkening of the auditorium, but also a kind of music that was itself noise. Printed piano score excerpts from Wagner's works, such as *Liebestod* and *The Ride of the Valkyries*, accompanied films long before *Apocalypse Now*, where *The Ride of the Valkyries* was no longer shown as a *lanterna magica* effect, as it was in Wagner's opera in 1876, but rather as a helicopter attack in the Vietnam War.

This brings us back to war and its innovations. In a word: World War I transformed Edison's simple light bulb into the electron tube, which made the live musical accompaniment of silent films obsolete. I am interested in the historical development of this technical wonder because the tube allowed for the possibility of synchronized film soundtracks and television up to the present day. It was not replaced until the development of contemporary LCD displays and other semiconductor technologies.

The electron tube, as I said, emerged from Edison's simple light bulb, which allows me to bring the history of lighting to a close. Edison had methodically searched for a cheap and safe light – so methodically that he brought every conceivable type of tropical wood to his laboratory as a possible filament for his bulb. The accidental combination, on which Daguerre had still subsisted, was thus systematically eradicated. Edison would have been able to electrify America after a couple years of research if a considerably more powerful competitor named Westinghouse had not replaced his direct current system with an alternating current system. On the other hand, Edison's discovery that light bulbs also work as electron tubes, as they emit ions under electrical voltage, was made entirely in passing. He was also unable to do anything more than have this so-called "Edison effect" named after him simply because he knew nothing about theoretical physics.

For this reason, a physics professor at the new and very modern *Reichsuniversität* in Strasbourg named Ferdinand Braun was the first to discover a possible application of the Edison effect in 1897. He deflected the electron beam inside the tube with electromagnets, which were in turn attached to the general alternating voltage of the Strasbourg power grid, and sent it to a phosphorescent screen. The controlled beam – the last and most precise variant of the actively armed eye – then inscribed the visible graphic sine wave of an alternating power supply on the screen. Braun had invented the oscilloscope. When his assistants later suggested to him that the electron beam should project beautiful images rather than mathematical functions, Braun rejected this first notion of television receiver tubes. He was "personally surprised" that Westinghouse's alternating power grid had not generated any ugly jagged peaks or rectangles, but rather an "ideal sine wave" (Kurylo, 1965, p. 137). Oscillograph means "vibration writer," and it is therefore the electronically perfected variant of all the movement writers, from Scott to Marey, that led to the writing of sounds and images. You will notice that the television played back equations rather than film characters when

191

it first began with Ferdinand Braun. It will possibly do so again at the end.

Braun's tube was not crucial for film and radio technology, however, but rather another tube variant: the so-called triode. Lee de Forest in Palo Alto and Robert von Lieben in Vienna simultaneously conceived the idea of building tubes out of two electrical circuits, one controlling and the other controlled. Two inputs were needed along with a general ground return, and it was therefore called a triode or three-way in the artificial Greek of technology. According to Pynchon's brilliant commentary, this separation of control circuit and output circuit in 1906 solved a fundamental problem of the twentieth century: that of control. Triodes were actually more bulky, they were more sensitive to heat, and they required more voltage than the transistors that have replaced them since 1947, but they were also unbeatably economical. In other words, a variably small control current, which assumed the function of Braun's electromagnets, could switch variably large output currents on or off, thus amplifying or weakening it. Thus, the electron tube first decoupled the concept of power from that of physical effort. But because power does not simply have negative effects, according to Foucault's thesis, the tube is also economically still insufficiently described. Immediately before the outbreak of World War I, de Forest discovered for the allies and Alexander Meißner for the central powers that tubes not only amplify but also provide a new type of power called feedback. When the output current of a triode is steered in the opposite direction of the control grid – when a voltage decrease in the first circuit thus corresponds to a voltage increase in the second circuit – negative feedback can be generated by leading the output signal, which for physical reasons is always delayed for fractions of a microsecond, back to the control circuit. The entire feedback system begins to oscillate between the minimum and maximum voltage without breaking up the oscillation. In other words, it becomes a high-frequency transmitter, which must then only be coupled with a low-frequency amplifying tube in order to send radio or television signals. In the first step, after they are converted to electricity, the acoustic or optical data signals are variably increased through low-frequency amplification. In the second step, the data signals become transmissible without wires over variable distances by modulating them on a high carrier frequency.

As the basic circuit of microphones and radio transmitters, this was already clear in 1913. But World War I provided new applications for the technology and the means for its mass production. Trenches brought an end to the possibility of commanding soldiers from a distance through optical or acoustic signals, yet this was precisely why a need for electronic feedback between invisible fronts and equally invisible control centers emerged. The first radio transmitters only served to entertain radio operators with music in exceptional cases, and eventually this was officially forbidden as a "misuse of army equipment." But to a greater degree they were used to manufacture feedback loops between ground personnel and reconnaissance flyers, who were told over the radio which enemy objects were invisible from the ground and still needed to be photographed or filmed.

This high-frequency military technology led to the worldwide explosion of electronics companies. Five years later, and again through the misuse of army equipment, the national radio institutions of Europe and the commercial stations of the USA emerged, and this was followed a decade later by the first television transmitter.

But low-frequency military technology also had consequences for entertainment electronics. The triumph of amplifier tubes allowed electronics companies to revolutionize Edison's and Berliner's old-fashioned mechanical sound recording technology. AT&T in the USA and Siemens & Halske in Germany wired a record player with a pick-up and electrically controlled speakers. This also resolved the problem that Edison's sound recording system failed to remedy in the Black Mary studio. The trumpet of the phonograph worked only when it was held directly in front of the actors' mouths, and it could thus embarrassingly be seen in the film that was being made at the same time. Tube amplifiers first made media acoustics into a sixth sense that could match up to the sixth sense called the camera.

Sound film was developed simultaneously in Germany and in the USA immediately after World War I, but it is completely senseless, at least on the American side, to list the individual inventors by name. It is enough to know that Warner Brothers was in serious financial trouble compared to the competition, and for this reason they reached for the life saver of sound film. The leading American electronics laboratory, AT&T's Bell Labs, gave the technically clueless Sam Warner a hand. After the record had been electrified, AT&T was also able to offer Warner Brothers a special model: a huge record that could be synchronized with the silent film and broadcast in the cinema hall using an amplifier and loudspeaker. This vitaphone system was and remained a patent of Western Electric.

For this reason, it was useless to interpret or explain films any more. Because the content of a medium is always another medium, sound films simply enhanced the reputation of the electronics companies that had made them possible. The first sound film in 1929 was

193

not called *The Jazz Singer* by chance. A Jewish ex-choirboy, who had sung with religious wise men in synagogues in New York as a child, defects to the American media after puberty and becomes a jazz singer (at least that is what white people called it in 1929). It breaks the heart of his pious father, who had not yet been disabused of his Mosaic faith by the melting pot of New York. The jazz singer is in the middle of a concert when he receives the news that his father is dying – let us say of heart failure – and he strikes up an old Jewish song that brings the entire audience to tears. Let us rather say: the recently electrified record companies, which are on the hunt behind the backs of all concert-goers, have a new hit in record sales. *The Jazz Singer* implies, therefore, that with the introduction of sound film Hollywood became a branch of electrical companies like Western Electric or General Electric, which possessed both record companies and radio stations at the same time and which, in turn, were only branches of large banks like Rockefeller or Morgan (Faulstich, 1979, p. 160).

In Germany, the development of sound film proceeded more systematically and on a much smaller scale. After losing the war, there was hardly any money, and instead demobilized army radio equipment stood around everywhere in 1919. In only four years, the signal corps had increased from 3,000 to almost 300,000 men. But even with misused army equipment, it was still possible to develop the very first sound film system without the use of records.

The developers of this wondrous work called it Tri-Ergon, which seems reminiscent of the triode tube but was actually intended to combine the names of the three developers – Hans Vogt, Joseph Masolle, and Dr. Joseph Engl – into a single anonymous "work of three." Luckily Vogt, the main player, left his memories to the German Museum in Munich. His "first contact with silent film" took place in 1905, when he was a 15-year-old peasant boy and he saw documentary films from the ongoing Russo-Japanese war in the cinematograph. "Eight years later," when he was already serving in the imperial navy at the experimental radiotelegraphy station in Kiel, the German auteur film had replaced the documentary cinematograph. Vogt enjoyed the "beautiful, highly dramatic" film *The Student of Prague*, as Evers and Seeber had just filmed it. However, there were two things that disturbed the young radio technician, who had been entrusted with the latest AEG tubes: "in close-ups, the lips of the actors moved like ghosts," and "the comments of the explainer ruined the atmosphere" (Vogt, 1964, p. 7). As is usual with autobiographies, Vogt claims that he would have immediately invented a new

194

media system if only World War I had not broken out. Four years later, partly at the front and partly at the high-frequency laboratory of a certain Dr. Seibt in Berlin, Vogt took part in the ether war:

> Soon I was active on the water, air, and earth fronts, soon again in the Berlin laboratory. A medium for communicating with buried trenches had to be created and tested. Radio direction finders and radio stations for the weather contingent zeppelins. The sad end of the war came. (Vogt, 1964, p. 7)

Vogt overcame this sadness while unemployed in postwar Berlin by once again returning to his film idea. With world war technology and know-how, it must have been possible to combine both of the media of the pre-war period – moving images without sound and constant noise without image – into the multimedia system of sound film. For this reason, the first thing that the Tri-Ergon people did was to establish technical specifications with systematic clarity:

> We take the following principles as the basis of our work:
> 1. The same film that carries the image must also serve as a sound carrier.
> 2. The sound must be recorded and reproduced through photographic processes.
> 3. All of the equipment necessary for sound recording, amplification, transmission, storage, reproduction and playback are not permitted to deform the original sound print. (Vogt, 1964, p. 11)

The most difficult part of this project was naturally amplification. Sound signals are initially so weak that at best they can make a hog's bristle vibrate, like Scott's phonautograph. Accordingly, the Tri-Ergon people first had to develop a tube amplifier that reacted "with a reproducible steepness of approximately six milliamperes of anode current change per volt of grid voltage change" (Vogt, 1964, p. 16). It turned out that a very similar tube amplifier had already been developed at Siemens by the great Dr. Schottky, to whom today's transistor technology owes all Schottky diodes. Patent rights thus no longer applied, but the Tri-Ergon people had still resolved their fundamental problem.

Only Tri-Ergon had now become a system project comparable perhaps only to Edison's electrification of theaters, streets, and residential homes. It could no longer be managed through individual inventions, but rather it required an entire chain reaction of new developments. After the solution of the amplifier problem, there was still the task on both the transmitter and the receiver side of transforming an acoustic signal into an optical signal that would be compatible with the filmstrip. Vogt, Massolle, and Engl solved this problem in a way that already prefigures the merging of sound and television technologies: because electricity had become the medium of all possible media or sensory channels, the sound signal could first be converted into current through a newly developed, highly sensitive, and inertia-free microphone, whose noise had been minimized compared to the old carbon button microphone. This current, which was still amplified by the new tube, then regulated a glow-discharge lamp, whose oscillations were visible and thus filmable when they were in the high-frequency range of up to 100 kHz (Vogt, 1964, p. 20). Despite their name, therefore, soundtracks are not sounds at all, but rather they are varyingly bright and varyingly wide images of the vibrations that sounds or noises physically are, which makes them extremely close to the Braunian tube.

And because the receiver side of a media system – according to Shannon's information theory – implements the inverse mathematical function of the transmitter side, the three Tri-Ergon developers constructed a selenium cell for their film projectors, which was also crucial for television. Selenium cells converted light into electricity again, which then in turn only had to be converted to the cinema sound system – and this was the historical reason why sound film engineers also included a few early television engineers, like Mihaly or Karolus. Vogt, Massolle, and Engl completed their technical specifications most elegantly: namely, they built the first electrostatic loudspeaker, which at that time could fill the entire cinema with sound and which is still ideal for headphones and stereos today. With this static loudspeaker, Edison's entire mechanics of sound storage was replaced by an electronic control.

So far, so good. Tri-Ergon had done its work and integrated an entire chain of new developments into a single media system. On February 22, 1920, the sound of a harmonica and the noteworthy word "milliampere" could be heard in a playing film for the first time (Vogt, 1964, p. 37). One year later, shortly after midnight, this contemptible yet wonderful word from the mouth of a technician was replaced with a female speaker "in close-up" reciting the poem *Heather Rose* by Johann Wolfgang von Goethe (Vogt, 1964, p. 38). This would have pleased Münsterberg, who claimed that film technically liquidates all classical-romantic substitute sensuality, such as the virtual reality of that rose, which as you know represents a virgin shortly before her defloration.

However, those sorts of successes did not yet make the new multimedia system a commercial success. As the Tri-Ergon people presented their technical *Gesamtkunstwerk* to the director of a large electrical company, which to my mind could only have been Siemens, their absolutely correct argument was that a systematically developed chain of tube amplifiers, microphones, and loudspeakers could support sound film as well as civilian wireless telephony – in today's words, therefore, entertainment radio. The director's counter-argument was that no listener "would spend good money for something that comes into his house for free like air and light" – in other words, no one would make Siemens happy by paying radio license fees (Vogt, 1964, p. 47).

In September 1922, the silent film industries reacted accordingly to the first public demonstration of sound film. From the viewpoint of their financial area of competence the *Berliner Börsenzeitung* briefly wrote: "The extent to which talking films are really the future, however, remains to be seen. It should not be forgotten that the talking film loses its internationality, and it must always remain limited to smaller works, as large films are only profitable on the world market" (Vogt, 1964, p. 44).

When Hans Vogt told this counter-argument to his wife, she came up with an idea that led to the Tri-Ergon people's single lucrative patent: the Gisela patent. Gisela Vogt proposed, namely, "to overcome speech difficulties in the future by making consecutive recordings of each sound film scene in the studio in multiple idioms, in the primary cultural languages" (Vogt, 1964, p. 44).

With this new principle of synchronization, which had been conceived by a woman, the multimedia system was perfect. By rights, it would have had to wipe out the cinema equipment that had already been established, but the three amateurs were not able to also finance this multi-billion dollar replacement. It was clear that the conversion of silent film to sound film was only the first in an entire series of conversions that would turn the existing media system operating worldwide completely inside out without interrupting its efficiency or its finances. The same holds true for television systems or technical wars in general, both in the recent past and in the future.

In the case of sound film, it is easy to predict the outcome of this immense need for capital: as in the USA, the German film industry also fell into the hands of electrical concerns like Siemens and AEG, which got involved in lawsuits with the American patent holders for years until the worldwide markets were divided up, as usual, and UFA was taken over by Deutsche Bank and Hugenberg. And one fine day, after the inconsequential Tri-Ergon patents had long been sold in Switzerland, an American film giant conceived of the idea of commercially surpassing Warner Brothers and their vitaphone system. William Fox, who had first made money, as I said, with Edison's automatic kinetoscopes, bought the Tri-Ergon patents and converted them into a completely auto-referential form of film publicity: Fox's Movietone talking newsreels, the first sound documentary film.

However, all of the international networks between companies and banks, or between Zurich and Hollywood, could not alter the fact that sound film – in contrast to silent film – does not constitute an international medium. The Gisela patent is and remains a compromise. As long as people continue to speak American or German, and have thus not yet defected to a worldwide standardized computer language like Algol, C or Ada, sound film will continue to serve as national propaganda in so-called national languages. This is rather inconsequential today, in the age of the computer, as the greatness of film is now a thing of the past and only computer languages still count. But because the English themselves refused at that time to recognize the dialect of Hollywood and California as English (Zglinicki, 1979, p. 612), sound film virtually appeared to form nations, just like the radio of that time. It thus restricted the companies of the interwar period to national language borders, which also committed them to the possibility of a second world war. Our film history must therefore turn to this war.

It should be emphasized beforehand that the difficulties of sound film synchronization also have an internal technical-aesthetic aspect. The numerous conversions between sound, image, and electricity that are necessary for this process already indicate that acoustic and optical signals are naturally less compatible. For precisely this reason, sound film was the first model case of a multimedia system long before television – if this term is understood as a system that manufactures not natural or physiological but rather technical connections. Sound film thus had to bridge two fundamental differences between optics and acoustics, which can briefly be explained.

First, acoustic signals only depend on a single independent variable: time (provided that stereophonic sound effects are disregarded, of course). Every musical signal is the momentary amplitude of a complex yet solitary vibration in time. As moving images, film, and television also depend on time. However, they also depend on two spatial coordinates: the height and width of each individual pixel (apart from 3-D experiments, the third spatial coordinate of depth

198

can be ignored in the same way that stereophonic sound can be disregarded when making sound recordings). A one-dimensional vibration or line is thus adequate for sound recording, whereas optical media principally require two-dimensional signal processing. When calculated by the technical-economic standards of engineers, this means that the amount of information to be processed is raised to the second power. For this reason, the image fills almost the entire celluloid of a sound film, whereas a band at the edge of the image is sufficient for the optical soundtrack, which in principle could even be implemented as a simple line.

Second, the acoustics compensate for this simplicity by functioning, since Edison's days, as an analog medium. In every given moment, the stored technical signal corresponds to the precise amplitude of the recorded sound signal. Until the introduction of audiotape, as I said before, this continuous image of continuous vibrations made it unimaginable and unfeasible to cut the soundtrack into pieces and then rearrange these clips arbitrarily, as is customary with film montage. According to the Breslau radio play pioneer Friedrich Bischoff, when the first radio play directors received the strange assignment to invent a new art form they in desperation chose the feature film as their specific model. Their quasi-cinematic radio play tricks even included slow dissolves and fades, which the new radio art borrowed from films. On the other hand, abrupt cuts and montages, which André Malraux called the operating principle of film, did not find their way into the emerging radio play.

By the same token, film and television are discrete processes that deal with nothing but individual frames or pixels, because optical storage is impossible even today. Even electronic switches work far too slowly to store or process light as the mixture of frequencies that it physically is. As is well known, audible sound ranges from approximately 25 Hz to 16 kHz, and that is why it was so easily conquered by low-frequency technology. Visible light is ten billion times faster and ranges from 700 to 380 terahertz. These are numbers with 14 zeroes, which is roughly 8 zeroes beyond the switching time of today's standard electronics, and therefore no medium could even come close. For this reason, film and television do not store light itself but rather only its photochemical effects (as we have already thoroughly discussed), which can then be stored and played back every twenty-fifth of a second, that is, in the low-frequency range. A compact disc that would sample sound events not at the usual 40,000 times per second but rather as slow as film or television would be a unique disaster for the ears.

The problems of television and sound film can only be understood on the basis of these physical principles of media technology. The Tri-Ergon process had to be capable of transporting the film images jerkily as a discrete signal, while simultaneously reading the optical soundtrack continuously as an analog signal.

To soften this collision between digital and analog, the Tri-Ergon people developed a new film transport system. Before and after a frame was projected, the celluloid went into a waiting loop, which compensated for the static images during the projection time. Behind the second waiting loop, the film roll continued running the entire time without any interruption, which also made it possible to read the soundtrack as an undisturbed continuum.

The mechanical solution to this problem also had a systematic side. Through millisecond-precise coupling between image and sound, sound film had carried out Messter's standardization proposals for the first time, which means that it had also required an absolutely fixed recording and playback speed for film images. Otherwise, a film star with a soprano voice would have been able to sound like a bass and a film star with a bass voice like a eunuch. You all know such sound manipulations from tape recorders that run too quickly or too slowly.

In the days of silent film, on the other hand, time loops and time lapses were not technical-experimental exceptions but rather daily practices in order to sell as many meters of celluloid as possible to a paying audience. For the first time, therefore, sound film opened up an absolute difference between experimental films and feature films. Real-time processing of the visuals first became genuinely verifiable through the ear and the acoustics. For this reason, sound film was a revolution in film aesthetics.

To begin with, many performing artists became unemployed in the middle of the global economic crisis. In 1929, a flyer with the following headlines in bold print was circulated in German moving picture theaters, as they were still so beautifully called (I have left out the fine print):

To the public!
Attention! Sound film is dangerous!
Sound film is kitsch!
Sound film is economic and spiritual murder!
Sound film is bad conservative theater at higher prices!
Therefore:
Demand good silent films!

Demand orchestral accompaniment with musicians!
Demand stage exhibitions with artists!
Reject sound film! (quoted in Greve et al., 1976, p. 287)

This social-revolutionary pamphlet, written in the style of the Communist Party or the Nazi Party, was of course signed by the *Internationale Artisten-Loge*, the performers union, and the *Deutscher Musiker-Verband*, the musicians union. These unions were somewhat justified, as the epoch of easy and therefore pointless theater or literary adaptations had indeed begun, but they were unable to comprehend a technology that entirely dispenses with acceptance or rejection, needs or fears. Circus artists and musicians were therefore just about as intelligent as many theorists, from Münsterberg to Béla Balázs, who praised silent film as a self-contained art form and harshly called the possibility of adding a soundtrack absurd. Only theorists, at least at that time, were less frequently unemployed.

The second point is reminiscent of a real historical joke that sound film played on a famous silent film star. John Gilbert had enchanted America's screens and women for 11 long years without having to betray the secret of his voice. However, in 1929 Gilbert's first sound film made what I have presented to you as a technical possibility literally or physiologically true: even without time-axis manipulation or sound time lapse, Gilbert's voice sounded like Mickey Mouse or a eunuch. Immediately after the première of this film, the star was a dead man in his own lifetime. Only Garbo threw a pair of roses into his grave (Zglinicki, 1979, p. 607).

The total securing of evidence through which people are registered by multimedia systems can also have a rebound effect on people themselves. Sound film changed the standard of voices and even more noticeably that of movements. As you all know, all of the expressionistic gestures employed by silent film actors for 30 years, which had barely made up for the lack of words and simulated the superimposed intertitles, disappeared; in facial expressions and gestures, sound film asserted a verifiable ordinariness. The crude time of small people – in other words, Hollywood cinema – could begin. For reasons that were less democratic than technical, it became practically impossible to cut or interrupt people in the middle of speaking, unlike the old days of silent film when the actors were inaudible. Even today, we must patiently watch and listen in front of the screen until no one in the studio has anything more to say. Expressionistic gestures, which had been developed as the last defense against cuts, film doubles, and trick film shocks, thus gave way to a new movement style, as the inability to cut acoustic events also dictated the optical events.

Funnily enough, one of the last German silent films, which had actually been made in the era of sound film, illustrated the difference between the two media within the plot as the difference between two generations. Using expressionistic silent film gestures, Fritz Lang's 1929 film *Frau im Mond* (Woman in the Moon) depicted an old, poverty-stricken professor who dreams only theoretically of moon rockets. In contrast, the young engineers, who turned this theory into blitzkrieg technology at practically the same time as the real Wernher von Braun, are depicted with the economical gestures of sound film, the new objectivity, and the *Wehrmacht*.

We can leave the content of sound films to themselves, because according to McLuhan's law the content of any medium is always only another medium. And thus, to adapt de Sade freely, we can bring the history of film with one final exertion to an end. With key words like rocket technology and *Wehrmacht*, a third subchapter has opened whose title, in a modification of a Foucault book, should be called: *From the Birth of Media Technologies to the Color of the World War*.

3.2.5   Color Film

In short, World War II was colorful. The reason was actually because the chief sponsor of the German film industry was Dr. Joseph Goebbels, the Minister for Popular Enlightenment and Propaganda, who had declared war on black-and-white film. In the name of total war or total simulation, World War II eliminated the last remaining differences between fiction and reality, and thus all the ways in which so-called artworks and so-called empiricism have been differentiated from time immemorial. After 1944, for example, German tank crews could no longer operate during the day due to allied air superiority, and they were equipped with night-vision devices that not only put Herschel's discovery of infrared light into practice, but also realized various dreams from a thousand and one nights. Conversely, Sir Watson-Watt provided the English air force with the first operational radar system in the world, which improved on Ritter's discovery of invisible ultraviolet light by extending it towards even higher light frequencies. If it was possible to see colors beyond even the visible color spectrum, such as infrared or radar waves, then visible colors were not permitted to remain hidden. The end of silent film as a consequence of World War I was thus followed by the development of color film as preparation for World War II. *The Redness of Red in Technicolor*, which was the lovely title of a 1960s book, was also the red over London, Dresden, and Hiroshima. The first attempt to use homemade Agfacolor colors was the film *Women Are Better Diplomats*, which was banned by Goebbels at the beginning of the war simply because in his eyes the colors were "shameful." The *Wehrmacht*, or more precisely the navy, had discovered feature films in American Technicolor on captured allied ships and sent them as spoils of war to the propaganda ministry. Among these films, naturally, was *Gone with the Wind*, the self-promotional color film about the American Civil War and the wonderfully colorful fires that develop when one burns the wooden palaces of old-fashioned southern slave owners. The German Agfacolor film followed this brilliant example and was gone with the wind. As usual, the leadership of the Reich contacted IG Farben, whose chemicals and poisonous substances were crucial to the war effort, and an optimized Agfacolor finally caught up with the criteria for realistic color set by Technicolor. This war over realistic color was supposed to define the entire history of optical media from 1939 to color television in 1965. But in 1942, Veit Harlan was already able to delight all of fortified Europe, from Cherbourg to Kiev, with his *Golden City*, a color film that was still primarily a form of publicity for itself. The propaganda minister's strategy, whereby only perfectly made entertainment could raise morale, had once again been put into action.

So much for Virilio's discussion of color film as the spoils of war (Virilio, 1989, p. 8). Incidentally, this was not only true for Veit Harlan, but also for Eisenstein, whose *Ivan the Terrible* had to make do with captured German Agfacolor film. The transfer of technological advances in two phases (from America to Europe and then from Europe to West Asia), which has since become typical, thus took shape long before Gorbachev.

In addition to these historical notes, however, some technical-systematic remarks must also be devoted to color film, and I will once again attempt to formulate this so generally that it will also be true for color television. This is easy from a technical perspective, because the black-and-white television from World War II already had the technical elements of color television.

I cannot compete with Goethe and deliver a historical theory of colors. Still, the most important steps leading from color photography to color film to color television must be mentioned. It has already been said that painters always had to deal with colors, and

203

signs of color deterioration over time became photography through positivization. But there was no economy of colors in painting, as this did not become necessary until the development of color ink and color photography. In other words, the number of colors that a painter used to produce a mixed color on his palette was entirely up to his discretion. However, in 1611, an entire century before Newton's analysis of the light spectrum, the Venetian Antonius de Dominis established the key principle that all colors can be mixed from red, green, and violet – the three primary colors. This additive synthesis made it possible to print copperplate engravings no longer only next to one another – as was already the custom with Gutenberg and his colleague Schöffler – but also over one another. Bold colors (on broadsheets) were replaced by infinitely graduated and perfectly overlapping color values, which were standardized in the late nineteenth century (after the discovery of chemical-artificial aniline dyes, the basis of IG Farben). With Senefelder's lithography, at the latest, three-color printing could commit all desired color nuances to paper. Four-color printing, which has since become the technical standard, also only uses three colors; the fourth printing plate is not colored, but is rather only black or grey, and its task is to differentiate brightness values (such as between pastel red and dark red of the same tint).

This economy of materials had important consequences for physiological optics, as it defined not arts but rather the first technical media. In the same years that Goethe spent dreaming of a poetic-aesthetic theory of colors, the Englishman Thomas Young developed first, his theory of the interference of light and second, his hypothesis that the eye also functions like a three-color print. Simply because it is completely improbable that the eye would have enough room to contain receptors for an infinite number of different colors, it was necessary to postulate the existence of sensors first for red, second for green, and third for violet. As far as I know, this has never been demonstrated on a living human eye, but in 1967 the Nobel Prize in Physiology went to three doctors who had proved that the cones of human-like vertebrates were limited to RGB (as the technicians say), while the rods can only distinguish between light and dark. Physiologically, therefore, every color signal is a mixture of different amounts of three tints plus a brightness or saturation value. Technologically, television – as electrified four-color printing – will derive from this its separation between chrominance (or color value) and luminance (or brightness value).

The first person to convert the new color physiology into technology again was none other than James Clark Maxwell, to whom we owe the general field theory of magnetism and electricity – the principle of all wireless radio communication – and the clarification that visible light does not consist of elastic vibrations, but rather it is only a special case of that electromagnetic spectrum. In 1861, Maxwell took black-and-white photographs of the same object, colored them in the three primary colors, and successfully projected the images over one another. This would later become the basic circuit of both color film and color television, and it had consequences for painters and technicians.

To begin correctly with the technicians: in 1868, the "ingenious [photographer] Ducos du Hauron" connected the physiology of Young (and Helmholtz) with the media technology of Maxwell (Bruch, 1967, p. 35). He literally suggested: "Small points and lines in the three primary colors red, green, and blue are placed next to each other on a plate so closely that they all merge into the same blended white at the same time. When the elements of all three colors are equally bright, they cover the same parts of the surface; when one color is less bright, more elements are taken from this color and they become larger" (Bruch, 1967, p. 35).

Ingenious blueprints rarely creep up so softly. First, Ducos du Hauron reversed the entire working principle of painting by replacing the subtractive color synthesis of a painter's palette, where the mixing of all colors together results in black, with Newton's and Maxwell's additive synthesis, where the sum of all colors results in white. Because there are no white phosphorescent materials in nature, incidentally, the white that has appeared on black-and-white screens since 1930 is also in technological reality a meticulously balanced mixture of complementary colors, whose addition results in this white. Second and possibly even more importantly, Ducos du Hauron began the digitization of color images. Just like the recently invented telegraph, he built colored areas out of "dots and dashes," which in the eye became the illusion of an apparently uniform colored area. Young's model, which broke the eyes down into nothing but discrete rods and cones for primary colors, was thus transferred to a picture or medium that could simulate human eyes and show them how to perform.

Painters only needed to transfer this technical model of vision to their canvases, through the mediation of a certain Chevreuil, and Europe's last object-oriented art movement was born. The so-called pointillists – Seurat, Signac, Pissaro, and so forth – actually built their landscape pictures out of nothing but points in the primary colors, which then within the eye, and indeed sensationally through additive synthesis, blurred into blended paints. This last competition between fine arts and optical media was followed only by abstract painting.

There was only one artist at the time who drew technical rather than aesthetic conclusions from these digitized colors: Charles Cros, who was a lyricist, painter, bohemian, and tinkerer all at the same time. He had already accurately described the phonograph shortly before Edison, and he also described the first method of making color photographs. After Young and Ducos du Hauron, all that remained necessary for color storage was the development of emulsions, which contain static, dispersed receptors for all three primary colors just like the eye. This is precisely what Cros proposed, although Edison said that he "was inconsistent, sought immortality at one time as poet and at other times strove for the painter's laurels, and thus he lacked that peculiar ability of concentration and inner composure, without which nothing great and permanent can be created" (Eder, 1978, p. 651). To put it more simply: the money from color photography was made by others.

In order to get to television, I will not bother tracing the lengthy history leading up to the mass production of Agfacolor in 1941. To put it briefly, it consisted simply in the commercialization of the theories and experiments we have already discussed. In the first phase, the rule already given concerning silent film still applied: silent black-and-white film was neither silent nor black-and-white. First, the technique of film tinting was developed on the basis of Maxwell's experiment: black-and-white films were simply dyed in uniform, monochromatic colors such that love scenes were depicted in pink, a yearning for nature was expressed through the color blue, etc. Second, there were attempts to color the thousands of frames in an entire film by hand, which was lavish, expensive, and rare – like impressionists painting the Cathedral of Chartres in all the changing shades of color of an entire day. Third, in 1897 the Berliner Hermann Issensee applied for a patent for color cinematography, which in the spirit of pointillism showed three differently colored frames one after the other in rapid succession, and these colors blended together at the extremely high frame rate of 120 hertz (Bruch, 1987, p. 19). Fourth, it is important to keep in mind that black-and-white films were never perfect before the development of panchromatic films, because the emulsion responded to the individual primary colors with varying degrees of intensity and these imbalances could only be adjusted through the use of expensive carbon arc lamps or sunlight (Monaco, 1977, p. 96).

In any case, you can see that it still took a long time after the introduction of sound film to erase the last difference between simulation and reality. As I said before, World War II was the first war in color. At the end of the war, the first color film material was provided to the propaganda companies that had been systematically established in all of the armies (see Barkhausen, 1982). In contrast to World War I, these propaganda companies were ordered to take part in the combat with weapons, and in contrast to Griffith or Messter they were consequently forbidden simply to reconstruct battle scenes. Unlike war correspondents on the radio, who were only equipped with transmitters, and unlike journalists, who were only equipped with typewriters (see Wedel, 1962), film could supply a multimedia show of color and sound, word and noise.

For this reason, the introduction of stereo sound and widescreen after the war was only a small step that enabled color film to be able to deceive the three-dimensionality of ears and eyes. In World War I, various professors, physicists, and musicologists had attempted to use perspective to detect invisible enemy artillery by artificially extending the range of vision or hearing. One of them, the French physicist Henri Chrétien, transferred his military detection technique to the civilian trick of horizontally compressing film images during recording to squeeze more optical information into Edison's standard format. During playback, however, this deformation would once again be equalized (Virilio, 1989, p. 69; Monaco, 1977, p. 87). Widescreen film was thus born, and it would be cinema's last life saver before the competition of television became overpowering. Up to now, the horizontal viewing angle of slipper cinema has been beneath contempt, but widescreen, cinemascope, and stereo sound can no longer change the fact that for us the highest purpose of film is already in the past. Strictly according to McLuhan's law, film has devolved into an evening program content filler for another medium: television.

3.3 Television

Unlike film, which simply inherited all the complexities of the image as accoutrements, television is a high-tech object. Therefore, we can and must link together many of the technical explanations that have already been given. It will be impossible, however, to draw any connections between television and literature or fantasies. Unlike film, there were no dreams of television prior to its development. In

1880, when the British humor magazine *Punch* published a caricature showing people watching television in the future, the principle underlying the technology had already been fixed. And when Liesegang edited his *Contributions to the Problem of Electrical Television* in 1899, thus naming the medium, the principle had already been converted into a basic circuit. Television was and is not a desire of so-called humans, but rather it is largely a civilian byproduct of military electronics. That much should be clear.

It should also be clear that the history of the development of television was the first realization through electronics of all of the functions named in Shannon's information theory. First, it was a fully electronic converter of images into currents, and thus a television signal source. Second, it was a fully electronic transmission circuit, and thus a television channel. Third, it was a fully electronic converter of current into images, and thus a television receiver. Its fourth function, which only developed much later, was also to serve as an electronic image storage device. The technical specifications were so complex because the digital processing of optical signals is two-dimensional; it must be able to cope with the square of the amount of information processed by records or radio transmitters.

Since 1840, however, the single electrical news channel was telegraphy, which had been standardized by Morse. It was a channel, therefore, that was just as linear as alphabet and letterpress – the media it had displaced. Just as letters are read one after the other, so too were the dots and dashes of Morse code transmitted one after the other through isolated copper-cored cables. On the other hand, Sömmerring's attempt to transmit the 26 letters and ten numbers over 36 parallel wires during the war of 1809 proved to be much too expensive and prone to interference. "A German idea," Napoleon reportedly said about Sömmerring's telegraphs. In contrast to film, therefore, the problem of television from the very beginning was how to make a single channel dimension from two image dimensions, and how to make a single time variable from convertible surfaces.

It was no coincidence that the principle answer was found by Alexander Bain, a Scottish philosopher and printer. Because the printing of writing processed data streams in a linear fashion, while the printing of images broke these surfaces down into dots, Bain was able to suggest that images should be principally conceived as rectangular grids, and that the individual raster elements should be transmitted point by point. In principle, therefore, images became discrete quantities of data, like telegrams. In strict opposition to photography, which prided itself on its analogy to nature, and in partial contrast to film, which consisted of a discontinuous or discrete sequence of many analog photographs, television began as radical cutting: it not only cut up movements in time, but it also disintegrated connections or shapes into individual points in space. The symbolic thus now triumphed where once the imaginary had ruled over perception and consequently also painting. And the fact that these individual points are today called pixels or "last elements" already shows how close television was to Young and Ducos du Hauron's theories concerning colored dots. This proximity had only to be realized.

It happened in 1883, even before the development of the feature film, when a 23-year-old physics student made a useable and patented television circuit based on Bain's principle. Paul Nipkow studied here in Berlin with Helmholtz, whose important experiments with sounds, voices, and colors laid the groundwork for the invention of the telephone and the gramophone. But while Helmholtz received an imperial physical-technical institute as a gift from Werner von Siemens, the telegraph industrialist, his student Nipkow remained in classic Mecklenburg poverty. In 1883, therefore, he spent Christmas Eve in his dormitory room in front of a small Christmas tree with the candles burning, a cheap petroleum lamp, and a German National Post Service telephone that one of his few friends had misappropriated as a private gift for himself. From this misuse of an imperial device emerged – on the last Christmas in the history of the world, if you will – not an entertainment medium, as our current misunderstanding might assume, but rather a channel in Claude Shannon's literal sense of the word. As it says in Nipkow's patent specifications, "the purpose of the apparatus described here is to make an object at location A visible at any location B" (Rings, 1962, p. 37).

The idea of this image transmission occurred to Nipkow either at the sight of the Christmas tree candles, which flicker, or at the sight of the telephone, which Alexander Graham Bell had first invented: if human voices were transmissible, why should it not also be possible to transmit the corresponding face? According to Bain, the image would then have to be cut up into individual points, which would be transmitted to a receiver using a telephone cable and would then be reassembled once again as a flickering image; as a good student of Helmholtz, however, Nipkow also knew about the inertia of the eye and its unconscious ability to filter out the image flicker either physiologically through the after-image effect already employed by film, or more generally or mathematically through the integration of individual pixels.

The sender of such a system stood (in his own words) "automatically" in Nipkow's mind's eye (if the literary quotation "mind's eye" had not become obsolete under the conditions of television). The so-called Nipkow disk – a metal disk that rotates around its axis – stood between the image source and the transmitting channel, and its sole function was to carry nothing but holes. When the disk turned, the holes arranged in spirals generated a visual axis of changing points on the image, which activated a fixed selenium cell that reacts to fluctuations of light with oscillating current, as already emphasized with sound film. The number of holes in the Nipkow disk corresponds to the desired number of lines in the television image, as Klaus Simmering points out:

> As the disk turns, one pixel after the other wanders in an approximately straight line over the screen, thus delineating an image line. If one point disappears on the left edge, the next point in the spiral, which shifts roughly one line space towards the center [of the disk], reaches the right edge and subsequently begins delineating the following line. The distance between the holes thus corresponds to the width of the screen, and the difference between the distance of the first and last holes from the center of the disk defines the height of the screen, which necessarily assumes a more or less trapezoidal form. (Simmering, 1989, p. 13)

To put it more simply: Nipkow imposed the discrete line form of a book or telegram onto images with sweeping success.

The inventor was less successful on the receiver side. What was missing was a method of converting the weak current produced by the photocell back into visible light, which in turn would have then been cut up by a Nipkow disk and projected as a two-dimensional image. Above all, Nipkow did not waste much thought on how to ensure that the continuous stream of electrical pixels on the receiver side would construct precisely the same lines as on the transmitter side. For example, if a 20-line television were reproduced with line 1 as line 19, line 2 as line 20, but line 3 as line 1, the imaginary of pattern recognition, on which Nipkow's video phone entirely depended, would have greatly suffered. In other words, the patent awarded by the imperial patent office in Berlin on January 6, 1884 did not provide a solution for a synchronization problem, and because this problem occurred between the transmitter and the receiver it was even more dramatic than the synchronization problem of image and sound with sound film.

Nipkow's third handicap was the mechanics of image scanning and image reconstruction. The electrical channel had the advantage of moving at the same absolute speed as the optical data itself; the selenium cell also kept pace, although it was admittedly slower than later Kerr cells (Simmering, 1989, p. 15); however, the rotating disks at both ends of the system were hopelessly slow and obsolete.

For this reason, Nipkow never thought about his Christmas idea again until his old age, when imperial German broadcasting dragged him once more into the spotlight of Berlin radio shows as a pioneer of German technology. One year after his patent, he joined the Berlin Railway Signal Company as a constructor, and for the next 30 years he constructed nothing but security equipment and signalling devices. He thus exchanged electrical high-frequency television signals, which were not yet technically feasible, for mechanical railway signals, and during the years of this regression he even forgot that he had applied for a television patent at all (Rings, 1962, p. 37).

Others fared no better than the patent holder. The processing of moving images in real time practically never succeeded with Nipkow disks; it only succeeded in later practice when real time played no role. In 1928, long before the Federal Criminal Police Office's computer surveillance, the Imperial Criminal Police were already able to slowly scan wanted posters and fingerprints, broadcast them with electrical speed, and then convert them slowly back into images again at distant police stations. Television in Germany thus did not begin with entertainment broadcasts; the states first learned how to secure evidence from across the country, which has led to remarkable arrests even quite recently.

But back to Nipkow's patent. I am sure you already anticipated that his three weak points were not corrected until the development of the tube. First, the tube enabled the virtually unlimited amplification of low-frequency currents, like those supplied by selenium cells after images are scanned. Second, the tube, at least in principle, made it possible to turn Nipkow's wire-bound video telephony into television, which like radio also transmits high-frequency wireless signals. By modulating the low-frequency pixel fluctuations, which cannot be radiated as such from any antenna, with a high carrier frequency, the television transmitter is complete. Third, the tube originated as the Braun tube even before it served the functions of amplification and feedback. Braun's oscilloscope could not only convert current immediately into light, thus avoiding all of the light bulbs that were still necessary for Nipkow, but it could also direct the luminous electron beam to arbitrary points on a screen using electromagnets, thus forming truly digital images out of nothing but points of light.

211

In 1897, Braun had made alternating current visible with his oscil-loscope, but he was not interested in transmitting images, as I have said. In 1908, on the other hand, Campbell-Swinton proposed the installation of a revolutionary television system with a Braun tube on the transmitter side as well as the receiver side. The image to be recorded was supposed to be projected on the screen of a Braun tube, which was covered with a mosaic of inertia-free photocells. The elec-tron beam of the tube was supposed to read out the electrical charge generated by the light falling on the photocells, thus converting it into oscillating current. On the receiver side, another Braun tube would then perform the reverse process by converting this current back into visible images (Simmering, 1989, p. 25). That is precisely how it happened.

But it took a long time to eliminate all of the mechanical weak points of Nipkow's transmission chain. Because every information system is only as good as its weakest component, the development of television was slowed down above all by the bandwidth of medium wave radio, which arose from World War I. These medium wave transmitters, in contrast to today's VHF, actually offered enough frequency bandwidth to transmit acoustic or one-dimensional signals, but a Braun television tube raised the amount of information by a power of two. To adapt Morgenstern freely, medium wave trans-mitters were not built for this at all, and it exploded the capacity of the channel. For this reason, mechanical Nipkow disks contin-ued to be used through the twenties even though the frame rate was a pathetic 12.5 hertz, which was still below that of film, and the 4 × 4 cm large image only had 30 lines. The German National Post Service experimented with this standard, and the engineers in charge were professors Mihaly and Karolus, who also worked on optical sound. The British Broadcasting Corporation (BBC) also experimented with this standard, with the passionate Scottish dilet-tante John Logie Baird in charge. For people, on the other hand, the blurredness of a scanned newspaper photograph magnified 20 times seemed meaningless; only a few amateurs, who had to synchronize their private Nipkow disks with the transmitter's Nipkow disk using fingerprints (Simmering, 1989, p. 17), played along with the new technicians' toy.

The thirties were thus influenced by a double optimization. First, propaganda experts had to invent images that would tune the popula-tion into television, which was best accomplished by the spectacular Berlin Olympics in 1936 and the Nuremberg rally crowd scenes. Second, even after broadcasting had electrified the television channel, technicians still had to replace the Nipkow disks with tubes. Manfred von Ardenne, who played a leading role in the history of the National Socialists and the German Democratic Republic, succeeded at playing back television images, and Vladimir Zworykin, an officer of the Tsar who emigrated to the USA, succeeded in recording them.

Strangely, it was once again at Christmas in 1931 that von Ardenne astonished journalists with his flying spot scanner, which had a resolution of 10,000 pixels and a frame rate of 16 hertz. Von Ardenne thus achieved an image quality that was far superior to that using mechanical Nipkow disks and even the frequencies of radio at that time. The channel capacity of Ardenne's television images increased even more after an ultra short wave radio was developed under pressure from the *Wehrmacht*, which was the only army in the world with ultra short wave radio-controlled tank divisions engaged in a blitzkrieg in 1939. As a BBC correspondent wrote, "Herr Eugen Hadamovsky, Director-General of the German broadcast service, [established] the world's first regular high-definition television service on Friday, 22 March 1935" (Simmering, 1989, p. 12).

But the German National Post Service's first regular high-definition television cannot be compared to Sony's high-definition TV today. It did not even have adequate transmitter tubes. The result was that only one other medium could be considered for the content of television: sound film, which had just recently been developed, was divided into image and sound. Radio transmitted the sound quite easily, while a Nipkow disk laboriously scanned the image – not even in real time at first. McLuhan's law was thus absolutely valid. Television could only be decoupled from feature films and turned into a live broadcast medium by developing an endless film: immediately after its development, it would be scanned by television, given a new emulsion and then it would be exposed and scanned again, etc., etc. This technique enabled the transmission of television in real time, just as radio broadcasters produced their own expensive canned recordings prior to the development of audiotape.

Third, the so-called "sweatbox" emerged as a genuine television recording studio, which functioned without the interposition of film technology. However, because the (even improved) Nipkow disks could only scan objects that were lit externally rather than self-illuminating objects like candles or light bulbs, the first television actors – like the telephone exchange after 1900, there were naturally no spokesmen but only spokeswomen – had to perform in an absolutely black box that was illuminated by the strongest available lamps, which was therefore also extremely hot. It was the last time in

213

history and the first time since Edison's Black Mary that the *camera obscura* was implemented.

All of these deficiencies on the transmitter side disappeared with the 1936 Olympics, a spectacle of world war sport staged by the National Socialists. A young engineer named Walter Bruch focused his electronic camera on the athletes from a specially constructed, invisible trench or television bunker in the Berlin Olympic Stadium. The iconoscope developed by Zworykin was actually supposed to be mounted in the heads of rockets to enable remote military reconnaissance (Virilio, 1989, p. 75); however, Zworykin's company, the Radio Corporation of America, turned it into a twin of von Ardenne's playback tube: the massless and inertia-free recording tube, just as Campbell-Swinton had imagined. Because Nipkow's individual selenium cells were replaced with an entire screen composed only of light-sensitive elements, the iconoscope supplied a light yield that was 40,000 times better, and it thus released early television stars from their sweatboxes.

With ultra short wave radio as the transmission channel, the iconoscope as the recording tube, and the flying spot scanner as the playback tube, the high-tech information system known as television was finally complete, because its combined functions (in Shannon's sense) shifted from mechanics to electronics. Like sound film, however, only countries and global companies on the technical warpath could still finance the fully electronic television system. The German National Post Service, for example, gave its patents on phosphorescent chemicals to the USA, as they were necessary for the apparent black and white of receiver tubes; in return, they received the iconoscope patented by RCA.

The political effects of this new image and sound medium were also similar to the effects of sound film. Television became a medium of national and domestic politics because it was transmitted in national languages and its extremely high transmission frequencies (prior to the development of satellite and cable television) only had a quasi-optical range of 60 to 70 kilometers. It was therefore no wonder that the Paul Nipkow television station, which had been named in honor of the inventor who was still living at that time, immediately had political functions. Hitler and Goebbels explicitly stated that novelists would be permitted to retain the completely ineffective medium of print provided that the state alone maintained a monopoly on all sounds and images. Despite its own claims or those made by its enemies, however, the National Socialist state was not monolithic and totalitarian but rather a conglomeration of power subsystems, so

214

television got involved in a war even before the beginning of World War II. After a long struggle between these power subsystems, from which the Third Reich emerged like a polyp, in December 1935 there was a decree: the National Post Service retained the rights to civilian technical developments, the Ministry of Information retained the rights to "representative forms for the purpose of popular enlightenment and propaganda," while the Air Ministry "in consideration of the special meaning of television for air traffic control and civil air defense" retained the rights to manufacture and distribute all television technologies (Bruch, 1967, p. 53). This tripartite division between civilian technology, program content, and military armament already proves that prior to World War II television was not a mass medium that would have derived its mass impact paradoxically from the intimacy of the picture size, program, and reception.

In fact, television worked much more like radio broadcasting, as dictators from Berlin to Moscow did not just rely on the intimacy of a recording microphone and a room speaker, but rather they rehearsed the mass impact of loudspeakers at party rallies or in Red Square. Following the model of the Volkswagen or the *Volksrundfunkempfänger* (people's radio), the electrical industry developed the *Volksfernseh-Einheitsempfänger E1* (people's television) at a price of 500 reichsmarks. It was the first rectangular tube in the world (Bruch, 1967, p. 71), and it stood in post offices and other public agencies in Berlin, where its screen was enlarged so that it could be seen by many spectators at the same time. The people who attended these "large picture stations" did not actually pay an entry fee, but they had to show tickets, as in the cinema, to regulate the amount of traffic. These broadcasts ran continuously with only brief interruptions from the outbreak of the war in 1939 until the bombing of nearly all the German transmitters: the few television receivers that were actually manufactured (50 instead of the planned 10,000) stood in military hospitals in Berlin and occupied Paris, where France's national television service could be connected directly to the *Wehrmacht* in 1944.

Now that we are discussing World War II, it is time to pause for a moment. It was clear since the Renaissance that perspective was closely related to firearms and ballistics. Photography was also applied to criminology and cryptography. World War I reconnaissance planes even connected film cameras to machine guns, and sound film was also developed on the basis of war technologies. But the high-tech medium of television is the only one among all of these optical media that functions according to its own principle as a weapon. For this reason, it would not have risen to world power without World War II.

We begin the overview of World War II television technologies in France, where a color television system named SECAM, based on *Wehrmacht* television, was promoted after the war. The developer of the SECAM system was Henri de France (1911–86), who worked in the field of radar and whose company received their commissions after 1930 "primarily from the War Ministry, the Department of the Navy and the Air Ministry. In 1934 Henri de France applied for his first patent for a direction finding system. Under contract with the French navy he developed a television system for the armoured cruisers of the second Atlantic fleet using a cathode ray tube and with an image resolution of 240 lines. In 1936 he succeeded in establishing a wireless television connection between warships on the high seas and the port of Brest" (Bruch, 1987, p. 63).

French postwar television was a product of radar – an electronically perfected variant of World War I detection methods – which had begun with analog or natural sensory media like the eye and the ear. Great Britain proceeded similarly, but much more systematically; unlike the *Wehrmacht*'s aggressive ultra short wave tank campaign, they had to prepare for a defensive war. The German physicist Christian Hülsmeyer had already successfully received the first electronic echo on May 18, 1904. He transmitted a radio impulse that traveled through space at light speed, was reflected by surfaces in its path, and was then once again received at the same location as the transmitter. When the signal delay was divided in half and multiplied by the speed of light, the result of course was the distance between the transmitter and the object that had reflected the signal. According to Virilio's brilliant formulation, therefore, radar is an invisible weapon that makes things visible (Virilio, 1989, p. 75) because it converts objects or enemies that do not want to be seen or measured at all into involuntary and compulsive transmitters (with the exception of the US Air Force's brand new stealth bomber). For the strategic benefit of Great Britain before the war, Sir Watson-Watt developed Hülsmeyer's basic circuit into a functional radar network. Radar stations were connected by radio throughout all of southern England, and they could report attacking Messerschmitts or Heinkels of the German *Luftwaffe* even while the approaching planes were still invisible. It was for precisely this reason that on the day the war began, the BBC discontinued the civilian television service it had introduced in 1936; from then on, the same high-frequency tubes that worked in television transmitters were sent to radar stations, and the same screens that worked in television receivers also made the invisible enemy visible on radar screens. Without any friction loss, an entertainment medium had been converted into a war technology. And because Watson-Watt understood that the quality of radar images was strategically crucial and inversely proportional to the wavelengths of the sent signals, Great Britain developed increasingly higher frequency tubes. At the same time, these tubes also had the advantage of becoming smaller and smaller, until finally entire radar installations could also be constructed on board airplanes. Using such UHF and VHF frequencies, which had been researched and made usable strictly for military purposes, civilian postwar television later became a world power. First, however, they endowed Royal Air Force fighter planes with electronic vision: airborne radar first made their blind enemies on the *Luftwaffe*'s side visible, but after 1943 it also made the rivers, streets, and cities of the empire visible, which were destroyed by the carpet bombing of fighter-supported long-range bombers.

At first, the *Luftwaffe* could only counter this terror bombing by linking radar and anti-aircraft searchlights to form the *Kammhu-berlinie*, which was named after a *Luftwaffe* general who was also a *Bundeswehr* hero. This is the final manifestation of the actively armed eye – the spotlights used on the Russian-Japanese front. In World War I, these spotlights already gave rise to those anti-aircraft searchlights that later shone in the company logo of Fox's Movietone talking newsreels and were misused by Albert Speer during the 1935 Nuremberg Nazi Party convention to create the first truly immaterial architecture. As Berlin burned in 1943, the same Speer wrote that he was "fascinated" by the "grandiose spectacle" of British bombers, German anti-aircraft searchlights, and crashing enemy planes, yet he neglected to add that his "dome of light" had already evoked this spectacle. In the *Wehrmacht*'s defense system, anti-aircraft searchlights (a visible weapon that made things visible) and radar (an invisible weapon that also made things visible) were thus parts of a feedback loop, and anti-aircraft searchlights as well as *Luftwaffe* planes were both directed towards their targets. In our terminology, therefore, radar would indeed have to be called an actively armed but electronic eye.

More effective feedback weapons remained in the developmental stages. Walter Bruch, who had operated the iconoscope at the Berlin Olympics and later developed the German color television system known as PAL, spent the war partly in Peenemünde and partly at Müggelsee. In Peenemünde, his two television cameras filmed the start of the first self-guided rockets and immediately transmitted these images by cable to a concrete bunker, where the engineers could remotely activate the V2 without being blown up by an eventual false start. Television thus assumed one of the basic functions of mathematical simulations: namely, using feedback loops to shield from something real.

At Müggelsee, the General Electricity Company (AEG) engineer succeeded in creating an even more promising feedback system: he left a pleasure steamer (in Bruch's words) "without any passengers of course" bobbing up and down on the lake for an approaching bomber to take aim at. This ballistic function was not assumed by the bomber pilot, however, but rather Bruch had constructed a television camera inside the bomb itself, which was supposed to be able to track down the enemy entirely on its own, follow it despite any evasive manoeuvres, and blow it up. World War II thus produced the first self-guided weapons systems, which have since made people, the subject of all modern philosophies, simply superfluous.

With the end of the subject, a television audience became possible in the postwar period, and with the triumph of radar, color television became possible as well. As the only country in the war that did not need to fear air attacks, the USA did not discontinue its development of television for the sake of radar. At the same time, the radar theory that emerged during the war was a key inspiration for the theory of digital signal processing in general. American physicists and mathematicians like Shannon were the first to come to the conclusion that telecommunications overall should not be based on continuous oscillations or waves, but rather on simple discrete radar impulses. There was a clear correlation between the precision of radar echoes and the wavelengths of transmitted signals: the shorter and steeper the signal, the more precise the echo. The rectangular pulse discovered through radar thus became fundamental for modern telephone networks, computer circuits, and even television standards. It was no wonder, therefore, that the USA emerged from World War II as the leading power in television technology. It was also no wonder that the war was continued by technical and economic means: it became a war over the standards of the worldwide mass medium television, which has not yet been resolved. When asked whether television was art, for example, Klaus Simmering answered: "Television is an internationally standardized way of seeing defined in CCIR Report 407–1" (Simmering, 1989, p. 3).

Unfortunately, there is no more time today to describe the war over television standards with all its victories and defeats. In terms of the theory, I can only remind you of the difference between styles and standards from the first lecture, and in terms of economics I can only remind you of the fact that when it comes to changing the standard of a high-tech medium billions of dollars are always at stake. As a brief history, it should finally be said that beginning in 1941 the USA introduced the leading black-and-white standard at that time, which still remains dominant today: 525 lines interlaced with a display of 30 frames per second instead of 25 frames per second, which later became the standard in Europe. This standard was developed simply because the power frequency in the USA is 60 hertz rather than 50 hertz and because it is important to synchronize the power grid and the frame rate to avoid optical noise. In an extremely short amount of time, this standard initiated the death of cinema and turned radio into a secondary medium. While President Roosevelt had still delivered fireside chats over the radio during the war, it is well known that John F. Kennedy defeated the cold warrior Richard Nixon in the presidential election of 1960 through a single television debate, in which he proved to be more telegenic.

In its competition with cinema, on the other hand, television still had much to learn, and it had to catch up with the war innovation of color film. The American network CBS made a first attempt at this, but naturally not until after the world market had been saturated with black-and-white televisions. Unfortunately, the Columbia Broadcasting System had learned nothing from a high-tech world war; it presented a color television even more primitive and mechanical than the Nipkow disk. An aperture with three sectors rotated in front of the screen, enabling the viewer to look at red, green, and blue frames one after the other. By the modest standards of American committees and populations, this was either too much or too little. The government and (as President Eisenhower's farewell speech about the "military industrial complex" had prophesied) the arms industry intervened, not only to create a better color television, but also to make black-and-white and color compatible. On the one hand, color television also had to be capable of being received on black-and-white screens, only without color. On the other hand, a color screen also had to be capable of correctly reproducing black-and-white broadcasts (Bruch, 1967, p. 91). To conform to these specifications, engineers from 30 electrical companies founded the NTSC or National Television Systems Committee. After 1954, the Federal Communications Commission, a central government agency that also controls the level of nudity and violence broadcast over the airwaves, made this committee the standard, and it subsequently became a big business. The only problem was that this

219

standard was oriented more towards economic profit than technical feasibility.

To see perfect colors, it would logically be necessary to transmit three times more information as with black-and-white television. Instead of simply transmitting black-and-white frames, in other words, it would be necessary to transmit red, green, and blue frames. However, there would then only be enough space left in the ether for a single television program for the entire broadcast area. For this reason, NTSC either had to reduce or compromise the color information. That was possible and also correct because the human eye contains fewer color receptors than movement receptors. NTSC therefore broadcasted only imprecise or narrow wave band color information and used the resulting free space for compatibility with black-and-white broadcasts. After the color signal was divided into luminance and chrominance, brightness and tint, black-and-white receivers could only use pure luminance, while color receivers also decoded chrominance. With 5 MHz bandwidth for luminance, only 1 MHz bandwidth for chrominance and in comparison an infinitesimally small bandwidth for the accompanying sound, the technicians of NTSC just succeeded in compressing complete color television programs into a VHF or UHF channel. In contrast to radio signals, therefore, television signals never corresponded to analog vibrations, but rather they were extremely complex assemblages. Like a spelled-out sentence, they were composed of various different elements and they adhered to the appropriate rules of syntax; you could even say they had their own electronic punctuation marks, which naturally consisted of synchronization signals.

However, the complex syntax of the NTSC signal did not get through to the receivers at all. As a result of phase shifts along the transmission path, the acronym NTSC was popularly known as "Never The Same Color." Due to the fact that they were not self-regulating, it was constantly necessary to readjust the tints by hand. Two European world war engineers, Henri de France in France and Walter Bruch in West Germany, set out to correct this flaw. They both kept the color stable using a classic trick of all telecommunications since Shannon: they did not immediately relay the received signal to a line on the television screen, but rather they first stored the line in an electronic buffer memory. With the reception of the next line one twenty-fifth of a second later, the stored signal and the new signal could then mathematically correct themselves so that the tonal values were finally stabilized. What eluded stabilization, of course, was the world market. Even today, the world of color television (a world that has only existed since the development of modern tele-communications) is divided between three different standards: NTSC for North America and Japan, SECAM for France and a vanishing Eastern bloc, and finally PAL for the laughing "rest of the television world" (Faulstich, 1979, p. 93).

The same competition also flared up around the storability of television images. In the first 30 years, there was no possible way of storing television images anywhere other than on Edison's old-fashioned film. It was not until after AEG and BASF developed a high-frequency and thus an extremely high-fidelity magnetic audio-tape during World War II, which set new quality standards for Greater German Radio, radio espionage and later also the field of computers, that it was also possible to conceive of an analog optical storage device. AMPEX produced the first professional videotape in the USA in 1958, shortly before BASF, which at least allowed insti-tutions to partly abandon film, which was the production standard at that time. But because the bandwidth of video so dramatically or rather quadratically exceeded the bandwidth of audio, video devices did not become truly mobile until the rise of Japan as the leading elec-tronic power. Sony's first video recorders were actually not designed for household use, but rather for the surveillance of shopping centers, prisons, and other centers of power, but through the misuse of army equipment users themselves also succeeded in mutating into television reporters and cutters. Television has since become a closed system that can process, store, and transmit data at the same time and thus allows every possible trick or manipulation, like film or music elec-tronics. And every video clip shows how far the tricks of music and optics have surpassed the speed of film. The pleasure afforded by this technology should not allow two things to be forgotten: the television always also remains a form of worldwide surveillance through spy satellites, and even as a closed information system it still represents a generalized assault on other optical media.

Before I discuss this notion of television as an assault on other optical media, I would like to make one additional point about so-called video art, which usually identifies itself as explicitly non-commercial television with explicitly bad image quality (although this bad image quality is almost perfectly suited to today's television stan-dard). Norbert Bolz recently found the only possible answer to the question of why video art presents images that are worse than those of television: the teacher of Nam June Paik, the world's leading video art installer, not to say artist, was a certain Karl Otto Götz, who was stationed in *Wehrmacht*-occupied Norway during World War II and who was ordered by his officers to investigate interference images on radar screens. To accomplish this goal, Götz recorded the rather noisy medium of the radar screen with the equally noisy medium of film, and he discovered something like form metamorphoses or structural progressions in this multiplied noise. Nam June Paik's video art, this aesthetic of interference that is deliberately inferior to the television standard, can thus once again be defined as a misuse of army equipment (Bolz, 1993, p. 164).

A closed electronic system like today's color television can hardly bear to be next to closed electrical-mechanical systems like film, especially when the image quality and the level of fascination associated with film exceeds that of television by a few decades. Marshall McLuhan described this difference in quality with the attributes hot and cool. Film is a hot medium because its widescreen illusions result in a decrease in the spectator's own activity, while television is a cool medium because it only supplies a moiré pattern comprised of pixels that the audience must first decode back into shapes again in an active and almost tactile way. As the analyst of a historical condition, McLuhan is absolutely right as always, but unfortunately he characterizes this distinction as a natural difference between both of these media. Apparently, even media theorists do not sufficiently realize that perceptible and aesthetic properties are always only dependent variables of technical feasibility, and they are therefore blown away by new technical developments. It is well known, for example, that tubes were replaced by tiny transistors in 1949, which in turn were replaced by integrated circuits in 1965. This simple space-saving silicon technology, which was originally developed for American intercontinental ballistic missiles, has since revolutionized all electronics, including entertainment electronics. Consequently, in the most recent escalation, television can join in the attack on all 35 mm film standards.

This began, unfortunately or naturally, neither in the USA nor in Europe, where companies like AT&T, Philips or Siemens have been resting on their old TV laurels until only recently. In Japan, on the other hand, a collaboration between Sony, the company that created Walkmans and video recorders, and Miti, the notorious Ministry for Technology and Industry, already set the new television standard a decade ago: High-Definition Television or HDTV. The explicit purpose of this development, which is already being employed by Japan's national television, was to abolish what McLuhan called the coolness of the medium and replace it with so-called telepresence. To begin with, telepresence means widening the practically square picture size so that it fills both eyes or at least engages them like a wide screen film, and the television thus loses its peep show character. Second, however, telepresence also means increasing the number of individual pixels beyond this growth in the picture size, thus considerably decreasing the necessary distance between the chair and the television. In other words, the eyes are permitted much closer to the screen without being bothered by moiré effects or violations of the sampling theorem, which were common up to now. The images falling on the retina occupy a considerably larger angle of vision, and telepresence can thus be described as an invasion or conquest of the retina through an artificial paradise.

This artificial paradise has aesthetic consequences, but its technical consequences are far more significant. According to Simmering's hypothesis, HDTV ensures above all that no longer only the faces of family members and politicians will fill the screen in close-up, unlike the current television standard. A simulated intimacy, which was simply the effect of technical handicaps, could be replaced with a total intimacy that is entirely like film. The aesthetic consequence would be a revolution in programs, but the technical consequence would be a competition to rival the current production standard of 35 mm film for films as well as professional television plays. Because the as yet undiscussed electronic processing of images is infinitely more effective and infinitely cheaper than film editing and film montage, this equalization would also mean the end of celluloid. Film would become the big screen projection of HDTV tapes, while television would become the close viewing of those same tapes. This would be a radical standardization and reduction of manufacturing costs, but it would also cost billions of dollars to replace all the television and film systems on the planet, which means that it will pass to the Japanese electronics industry. None of the optical media standards up to now would satisfy the requirements of HDTV. It is precisely for this reason that the system will defy all European and American opposition, and it is precisely for this reason that I have warned the film and television scholars among you from the very beginning not to pin your occupational hopes on celluloid.

The aesthetic of HDTV is therefore clear, but the technology poses nothing but problems. A single HDTV transmitter with 1,125 lines, a frame rate of 60 Hz, a luminance signal of 20 MHz, a chrominance signal of 7 MHz and CD quality stereo sound requires a channel capacity, as you can easily calculate, of roughly 30 MHz. In other words, even under the conditions of UHF and VHF, this single transmitter would use the entire frequency spectrum of its reception area. However, because Japanese companies like Sony are not a direct continuation of the Greater East Asia Co-Prosperity Sphere, as the Japanese empire built them up to be during World War II, they show democratic mercy. They compromise the HDTV signal before it is emitted through a mathematical algorithm that is called MUSE of all things. When you hear this word, though, please don't think of the Greek muses of poetry, music, history, and the arts in general, which have been overcome, but rather think about the sampling theorem created by the AT&T engineers Nyquist and Shannon. The acronym MUSE stands for Multiple Sub-Nyquist Encoding (Simmering, 1989, p. 76), and using mathematical tricks it reduces the television channel bandwidth from 30 MHz to a tolerable 7 MHz. Sony's muse thus enables the broadcast of HDTV programs from conventional radio transmitters without limiting each reception area to just a single transmitter. If this high-tech muse did not exist, the only other remaining possibility would be a return from wireless transmission back to cable, as telegraphy was once defined. By now, mind you, these cables have become more advanced with the development of higher-frequency optical fibers. As you know, fiber-optic cables operate on the basis of laser light, which is reflected inconceivably often from one end of an inconceivably fine mirrored tube to the other. They thus represent the first and probably very significant method of exceeding the speed of electricity, which is considerably delayed by conductors. For the first time in the entire history of media, in other words, fiber-optic cables transmit optical signals as light rather than electricity, which enables them to absorb the enormous frequency band of HDTV. This sensational tautology of light becoming a transmission medium for light includes rather than excludes the possibility that the same speed of light also benefits all other signals. Besides television signals, optical fibers can also transport electronically converted acoustics, texts or computer data, and can thus be promoted to the position of a general medium, just as Hegel had already celebrated light.

— 4 —

COMPUTERS

Projects like ISDN (an integrated fiber-optic network for any type of information), which have long been in the planning stages, change not only the transmission methods of contemporary media systems but also the processing itself. The introduction of HDTV and ISDN conflates television not only with old-fashioned film, but also and above all with the medium of all media: computer systems. It is already clear that a data compression technology like MUSE is no longer concerned with genuine optical processes like signs, colors, and etchings (to formulate it in old-fashioned painter terminology). On the contrary, MUSE entails the application of rules for computing or algorithms on optical signals, which could be applied just as well in acoustics or cryptography because they are perfectly indifferent towards medial contents or sensory fields, and because all of them end up in that universal discrete machine invented by Alan Mathison Turing in 1936, the computer. In 1943, the computer had a mission that was crucial to the war: to crack the *Wehrmacht*'s entire coded ultra short wave radio. Ever since the Pax Americana has become the worldwide basis of all high technology, it has assumed the task of decoupling the knowledge of this planet from its populations and thus also making it transmissible on an interstellar level. For this reason, visible optics must disappear into a black hole of circuits at the end of these lectures on optical media.

To begin with, computer technology simply means being serious about the digital principle. What are only the edits between frames in film or the holes in the Nipkow disks or shadow mask screens in television become the be all and end all of digital signal processing. There are no longer any differences between individual media or sensory fields: if digital computers send out sounds or images, whether to

225

a so-called human-machine interface or not, they internally work only with endless strings of bits, which are represented by electrical voltage. Every individual sound or pixel must then actually be constructed out of countless elements, but when these bits are processed quickly enough, as the mathematician John von Neumann recognized in the face of his first atomic bomb, everything that is switchable also becomes feasible. The standard currently lies somewhere between ten and 70 million bit operations per second, but in the near future optical circuits could increase this number still further by a factor of a few million. In any case, Méliès' *Charcuterie méchanique*, with its cutting between frames every second and its cutting of a pig every minute, is now obsolete because a computer that processes or outputs audiovisual data functions like a cutter that no longer circumvents only our perception time (like all analog media), but also the time of so-called thinking. That is why every possible way of manipulating data is at its disposal.

In contrast to film, television was already no longer optics. It is possible to hold a film reel up to the sun and see what every frame shows. It is possible to intercept television signals, but not to look at them, because they only exist as electronic signals. The eyes can only access these signals at the beginning and end of the transmission chain, in the studio and on the screen. Digital image processing thus ultimately represents the liquidation of this last remainder of the imaginary.

The reason is simple: computers, as they have existed since World War II, are not designed for image-processing at all. On the contrary, it is possible to grasp the history of their development in connection with Vilém Flusser's notion of the virtual abolition of all dimensions. In Flusser's model, the first symbolic act, which began at some point in the prehistory of human civilization, was to abstract a three-dimensional sign out of the four-dimensional continuum of space and time. This sign stood for the continuum, but because of this dimensional reduction it could also be manipulated. Some examples are obelisks, gravestones, and pyramids. The second step consisted in signifying this three-dimensional sign through a two-dimensional sign. A gravestone could be signified by a painting of a pietà, for example, which once again increased the possibilities of manipulation. The third step was the replacement or denotation of the two-dimensional through the alleged one-dimensionality of text or print, which McLuhan's media theory also claims, although all of our book pages since the eleventh century are structured surfaces – but that deserves its own lecture.

What all of these reductions had in common was that the *n-1* dimensional signifier at the same time also concealed, disguised, and distorted the signified, that is, *n* dimensional. This is the reason for the polemics of Greek philosophers against gods of flesh and blood, the wars of iconoclasts or reformers against religious images, and finally, in the modern era, the war of technology and natural science against a textual concept of reality. In this last war, according to Flusser, one-dimensional texts have been replaced by zero-dimensional numbers or bits – and the point is that zero dimensions do not include any danger of concealment whatsoever.

When seen from this perspective, computers represent the successful reduction of all dimensions to zero. This is also the reason why their input and output consisted of stark columns of numbers for the first ten years after 1943. Operating systems like UNIX introduced the first one-dimensional command lines in the sixties, which were then replaced by a graphic or two-dimensional user interface in the seventies, beginning with the Apple Macintosh. The reason for this dimensional growth was not the search for visual realism, but rather its purpose was to open up the total programmability of Turing machines at least partially to the users, which demands as many dimensions as possible due to the inconceivable number of programming possibilities.

The transition to three-dimensional user interfaces (or even four-dimensional ones if time is included as a parameter), which today goes by the phrase "virtual reality," can of course also be understood as an expansion of the operational possibilities. Virtual realities allow for the literal immersion of at least two distant senses, the eye and the ear, and at some point they will also enable the immersion of all five senses. Historically, however, they did not originate from the immanence of the development of the computer, but rather from film and television.

An American named Fred Waller already realized in the thirties that normal film formats do not fill up the field of vision of two eyes at all. For this reason, Waller developed Cinerama, which combined three or even five cameras and projectors arranged next to each other. The films were projected onto semicircular screens, which surrounded the spectator so that the spectator's entire field of vision was consequently immersed in the film image. This technology was primarily designed for flight simulators, and it thus served a military purpose. In the fifties, Morton L. Heilig replaced Waller's film projectors with small television cameras directly in front of both eyes, which thus replaced the mass consumers in the cinema hall with a simple,

227

lonely cybernaut. Virtual reality as the bombardment of the senses, and above all of the senses of bomber pilot trainees, was born (see Halbach, 1993).

However, this hunt for visual realism should not deceive us with regard to the basic principles of computer graphics. The fundamental difference between Heilig and today's virtual realities is that Cinerama simply filmed the New York Broadway, while computers must calculate all optical or acoustic data on their own precisely because they are born dimensionless and thus imageless. For this reason, images on computer monitors – and there are now already almost as many as televisions – do not reproduce any extant things, surfaces, or spaces at all. They emerge on the surface of the monitor through the application of mathematical systems of equations. In contrast to television images, which ever since Nipkow's disks consist of more or less continuous lines but discrete columns, this surface is composed from the outset of a square matrix of individual points or even pixels, and it is therefore also discretely controlled on the horizontal axis. With super VGA, the leading monitor standard at the moment, the manic cutter known as the computer has free reign over 640 times 480 pixels and 256 different colors, and these variables are determined at the leisure of the image-processing algorithms. Whether the screen is supposed to represent the quantity of real numbers or complex numbers is mathematically only a question of convention. In any case, the computer functions not merely as an improved typewriter for secretaries, who are permitted to relinquish their old-fashioned typewriters, but rather as a general interface between systems of equations and sensory perception – not to say nature. In 1980, the mathematician Benoît Mandelbrot proceeded to analyze a very elementary equation of a complex variable point for point on the computer screen. The equation itself had been known since 1917, but it would take mathematicians at best millions of days to calculate it with paper and pencil. It is also significant that the color samples first made possible on the computer screen have since been given splendid names like "apple men," "cantor dust," or "seahorse region," as they produced a nature that no human eye had previously recognized as a category: the category of clouds and sea waves, of sponges and shorelines. Digital image-processing coincides with the real, therefore, precisely because it does not want to be a reproduction like the conventional arts. Silicon chips, which consist of the same element as every pebble on the wayside, calculate and reproduce symbolic structures as digitizations of the real.

For this reason, the transition from today's system, which consists of silicon chips for processing and storage and gold wires or copper webs for transmission, to systems of fiber-optic cables and optical circuits will exponentially increase not only the calculation speed of digital images, but also the mathematical structure of self-similarity discovered by Mandelbrot. For example, when a glass diffracts incidental light, producing the effects known since Fresnel as interference and color moiré, it is already by nature a mathematical analysis that could only be processed in an extremely time-consuming way by serial Von Neumann computers. So why spend so much effort translating this light into electrical information and then processing this information serially or consecutively if the same light can already calculate itself and above all simultaneously? At the end of this lecture, I would like to look ahead to the future of optical media, to a system that not only transmits but also stores and processes light as light. In a last dramatic peripeteia of its deeds and sufferings, this light will thus cease to be continuous electromagnetic waves. On the contrary, to adapt Newton freely, it will again function in its twin nature as particles in order to be equally as universal, equally as discrete, and equally as manipulable as today's computers. The optimum of such manipulability in the virtual vacuum of interstellar space is already mathematically certain. With this optimum, every individual bit of information corresponds to an individual light pixel, yet these pixels no longer consist of countless phosphorescent molecules, as on television and computer screens, but rather of a single light quantum or photon. Whereupon the maximum transmission rate of the information of a simple equation, which can no longer be physically surpassed, is: $C = (3.7007)(\sqrt{p/h})$. To put it into words, the maximum transmission rate of light as information or information as light is equal to the square root of the quotient of photon energy divided by Plank's constant multiplied by an empirical coefficient.

Equations are there for the purpose of being inconceivable and thus simply circumventing optical media and lectures about them. For this reason, allow me a single illustration at the end. Imagine an individual photon in a vacuum like the first star in the evening sky, which is otherwise empty and infinite. Think of the emergence of this single star in a fraction of a second as the only information that counts. And listen to this passage from Pynchon's great world war novel, where the old rocket officer from Peenemünde talks to the young man whom he sent on the first rocket trip into space, from which he will never return:

The edge of evening . . . the long curve of people all wishing on the first star . . . Always remember those men and women along the thousands of miles of land and sea. The true moment of shadow is the moment in which you see the point of light in the sky. The single point, and the shadow that has just gathered you in its sweep . . . Always remember. (Pynchon, 1973, pp. 759–60)

So much for the algorithms of random, namely digital data in the domain of images. What I have been able to tell you are only the algorithms that America's National Security Agency, the NSA, have released up to now. There are possibly algorithms from general staffs or secret services that have long been more efficient, but which are still top secret. It is impossible to persuade oneself that November 9, 1989 (the fall of the Berlin Wall) marked the end of every war. The east is surely defeated – through propaganda television at the consumer level and through computer export embargoes at the production level; but in the southern hemisphere there still remains the problem of information versus energy, algorithms versus resources, which is at least 200 years old.

In the world war between algorithms and resources, the 2,000-year-old war between algorithms and alphabets and between numbers and letters has practically faded into obscurity. For this reason, I would like to address my final words directly to you. For the past 14 lectures about optical media I have resisted the temptation to write my own computer graphics programs (whatever "own" means in the world of algorithms). Instead, simple boring lecture manuscripts emerged under the dictates of a text-processing program named WORD 5.0. As long as Europe's universities have not installed high-performance data lines to all auditoriums and dormitories, no other choice remains. Under high-tech auspices, however, the entire lecture has been a waste of time. I am comforted by the hope that your generation will lay the high-frequency fiber-optic cables and crack the secret world war algorithms. All that remains is for me to thank your old-fashioned open ears and to conclude with an old-fashioned rock song, which penetrated the ears of my generation, which as you know, nothing and no one can close.

Leonard Cohen, *A Bunch of Lonesome Heroes*

I sing this for the army,
I sing this for your children
And for all who do not need me.

BIBLIOGRAPHY

Alberti, Leon Battista. *On Painting*. Trans. John R. Spencer. New Haven: Yale University Press, 1966.

Arnheim, Rudolf. *Kritiken und Aufsätze zum Film*. Ed. Helmut Diederichs. Munich: Carl Hanser, 1977.

Bachmann, Ingeborg. *Songs in Flight: The Collected Poems of Ingeborg Bachmann*. Trans. Peter Filkins. New York: Marsilio Publishers, 1994.

Barkhausen, Hans. *Filmpropaganda für Deutschland im Ersten und Zweiten Weltkrieg*. Hildesheim: Olms Press, 1982.

Barthes, Roland. *On Racine*. Trans. Richard Howard. New York: Hill and Wang, 1964.

Battisti, Eugenio. *Filippo Brunelleschi: The Complete Work*. Trans. Robert Erich Wolf. New York: Rizzoli, 1981.

Belting, Hans. *Likeness and Presence: A History of the Image before the Era of Art*. Trans. Edmund Jephcott. Chicago: University of Chicago Press, 1994.

Benjamin, Walter. "The Work of Art in the Age of Mechanical Reproduction." *Illuminations*. Trans. Harry Zohn. New York: Schocken Books, 1969, pp. 217–51.

Bergk, Johann Adam. *Die Kunst, Bücher zu lesen, nebst Bemerkungen über Schriften und Schriftsteller*. Jena: Hempel, 1799.

"Bertillonsches System." *Meyers Großes Konversations-Lexikon*. 20 vols. Leipzig: Bibliographisches Institut, 1902–08. Vol. 2, 1905, pp. 732–3.

Bidermann, Jakob. *Cenodoxus*. Stuttgart: Philipp Reclam, 1965.

Blumenberg, Hans. *The Legitimacy of the Modern Age*. Trans. Robert M. Wallace. Cambridge, MA: MIT Press, 1983.

Boltzmann, Ludwig. *Populäre Schriften*. Ed. Engelbert Broda. Braunschweig/Wiesbaden: Friedrich Vieweg & Sohn, 1979.

231

Bolz, Norbert. *Am Ende der Gutenberg-Galaxis: Die neuen Kommunikationsverhältnisse.* Munich: Wilhelm Fink Verlag, 1993.

Bosse, Heinrich. *Autorschaft ist Werkherrschaft: Über die Entstehung des Urheberrechts aus dem Geist der Goethezeit.* Paderborn: Ferdinand Schöningh, 1981.

Braunmühl, Anton von. *Vorlesungen über Geschichte der Trigonometrie. 1. Teil: Von den ältesten Zeiten bis zur Erfindung der Logarithmen.* Leipzig: B. G. Teubner, 1900.

Bruch, Walter. *Kleine Geschichte des deutschen Fernsehens.* Berlin: Hande & Spender, 1967.

Bruch, Walter and Riedel, Heide. *PAL – das Farbfernsehen.* Berlin: Deutsches Rundfunk-Museum, 1987.

Büchner, Georg. "Leonce and Lena." Trans. Anthony Meech. *The Complete Plays.* Ed. Michael Patterson. London: Methuen, 1987, pp. 113–46.

Büchner, Georg. *Leben, Werk, Zeit: Ausstellung zum 150. Jahrestag des "Hessischen Landboten".* Marburg: Jonas Verlag, 1985.

Buddemeier, Heinz. *Panorama, Diorama, Photographie: Entstehung und Wirkung neuer Medien im 19. Jahrhundert.* Munich: Wilhelm Fink Verlag, 1970.

Busch, Bernd. *Belichtete Welt: Eine Wahrnehmungsgeschichte der Fotografie.* Munich: Carl Hanser, 1995.

Clark, Ronald William. *Edison: The Man Who Made the Future.* New York: Putnam, 1977.

Crary, Jonathan. *Techniques of the Observer: On Vision and Modernity in the Nineteenth Century.* Cambridge, MA: MIT Press, 1991.

Eder, Josef Maria. *History of Photography.* Trans. Edward Epstean. New York: Dover, 1978.

Edgerton, Samuel, Jr. *The Heritage of Giotto's Geometry: Art and Science on the Eve of the Scientific Revolution.* Ithaca: Cornell University Press, 1991.

Edgerton, Samuel, Jr. *The Renaissance Rediscovery of Linear Perspective.* New York: Basic Books, 1975.

Eisner, Lotte H. *The Haunted Screen: Expressionism in the German Cinema and the Influence of Max Reinhardt.* Trans. Roger Greaves. London: Secker and Warburg, 1973.

Enzensberger, Hans Magnus. *Mausoleum: Thirty-seven Ballads from the History of Progress.* Trans. Joachim Neugroschel. New York: Urizen, 1976.

Faulstich, Werner, ed. *Kritische Stichworte zur Medienwissenschaft.* Munich: Wilhelm Fink Verlag, 1979.

Flaubert, Gustave. *Sentimental Education.* Trans. Robert Baldick. London: Penguin, 1964.

Freud, Sigmund. *The Standard Edition of the Complete Psychological Works of Sigmund Freud.* 24 vols. Trans. James Strachey. London: Hogarth Press, 1953–74.

Gréard, M.O. *Jean-Louis-Ernest Meissonier, ses souvenirs – ses entretiens précédés d'une étude sur sa vie et son oeuvre.* Paris: Librairie Hachette, 1897.

Greve, Ludwig, Pehle, Margot and Westhoff, Heide, eds. *Hätte ich das Kino! Die Schriftsteller und der Stummfilm.* Marbach: Deutsche Schillergesellschaft, 1976.

Halbach, Wulf R. "Reality Engines." *Computer als Medium.* Ed. Norbert Bolz, Friedrich Kittler, and Christoph Tholen. Munich: Wilhelm Fink Verlag, 1993, pp. 231–44.

Heidegger, Martin. "The Age of the World Picture." *The Question Concerning Technology and Other Essays.* Trans. William Lovitt. London: Garland, 1977, pp. 115–54.

Herr, Michael. *Dispatches.* London: Picador, 1978.

Hoffmann, E.T.A. *The Devil's Elixirs.* Trans. Ronald Taylor. London: John Calder, 1963.

Hoffmann, E.T.A. *Eight Tales of Hoffmann.* Trans. J. M. Cohen. London: Pan Books, 1952.

Hoffmann, E.T.A. *Fantasie- und Nachtstücke.* Ed. Walter Müller-Seidel. Munich: Winkler Verlag, 1967.

Holmes, Oliver Wendell. "The Stereoscope and the Stereograph." *The Atlantic Monthly* 3 (1859): 738–48.

Innis, Harold. *Empire and Communications.* London: Oxford University Press, 1950.

Jauss, Hans Robert. "Nachahmungsprinzip und Wirklichkeitsbegriff in der Theorie des Romans von Diderot bis Stendhal." *Nachahmung und Illusion.* Ed. Jauss. 2nd edn. Munich: Eidos Verlag, 1969, pp. 157–78.

Jay, Paul. *Lumières et images, la photographie: Histoire sommaire des techniques photographiques au XIXe siècle.* Chalon-sur-Saône: Musée Nicéphore Niépce, 1981.

Kaes, Anton, ed. *Die Kino-Debatte: Texte zum Verhältnis von Literatur und Film 1909–1929.* Tübingen: Niemeyer, 1978.

Kant, Immanuel. *Critique of Judgment.* Trans. J. H. Bernard. New York: Hafner, 1951.

Kroker, Arthur. *Technology and the Canadian Mind: Innis/McLuhan/ Grant*. New York: St. Martin's Press, 1984.

Kurzel-Runtscheiner, Erich. *Franz Freiherr von Uchatius*. Vienna: Österreichischer Forschungsinstitut für Geschichte der Technik, 1937.

Lacan, Jacques. *Écrits: A Selection*. Trans. Bruce Fink. New York: Norton, 2002.

Lacan, Jacques. *The Four Fundamental Concepts of Psycho-Analysis*. Ed. Jacques-Alain Miller. Trans. Alan Sheridan. New York: Norton, 1981.

Lambert, Johann Heinrich. *Neues Organon oder Gedanken über die Erforschung und Bezeichnung des Wahren und dessen Unterscheidung vom Irrtum und Schein*. 3 vols. Ed. Günter Schenk. Berlin: Akademie Verlag, 1990, vol. 2.

Lasswitz, Kurd. *Gustav Theodor Fechner*. 3rd edn. Stuttgart: Frommann, 1910.

Lessing, Gotthold Ephraim. *Laocoön: An Essay on the Limits of Painting and Poetry*. Trans. Edward Allen McCormick. Baltimore: Johns Hopkins University Press, 1984.

Loyola, Ignacio de. *The Spiritual Exercises of Saint Ignatius*. Trans. Thomas Corbishley. London: Burns and Oates, 1963.

Mallarmé, Stéphane. *Œuvres complètes*. Ed. Henri Mondor and G. Jean-Aubry. Paris: Gallimard, 1945.

McLuhan, Marshall. *Understanding Media: The Extensions of Man*. New York: McGraw-Hill, 1964.

Monaco, James. *How to Read a Film: The Art, Technology, Language, History and Theory of Film and Media*. New York: Oxford University Press, 1977.

Morin, Edgar. *The Cinema, or The Imaginary Man*. Trans. Lorraine Mortimer. Minneapolis: University of Minnesota Press, 2005.

Münsterberg, Hugo. *The Film: A Psychological Study; The Silent Photoplay in 1916*. New York: Dover, 1970.

Nabokov, Vladimir. *Lolita*. New York: G. P. Putnam's Sons, 1958.

Nadar (Félix Tournachon). *Quand j'étais photographe*. Paris: E. Flammarion, 1899.

Nietzsche, Friedrich. "On the Future of Our Educational Institutions." *The Complete Works of Friedrich Nietzsche*. Ed. Oscar Levy. Trans, J. M. Kennedy. 18 vols. New York: Russell & Russell, 1964. Vol. 3.

Ong, Walter J. *Orality and Literacy: The Technologizing of the Word*. London: Routledge, 1991.

Pinthus, Kurt, ed. *Das Kinobuch*. Zürich: Arche, 1963.

Plumpe, Gerhard. *Der tote Blick: Zum Diskurs der Photographie in der Zeit des Realismus.* Munich: Wilhelm Fink Verlag, 1990.

Poe, Edgar Allan. "The Oval Portrait." *The Complete Works of Edgar Allan Poe.* Ed. James A. Harrison. New York: AMS Press, 1965. Vol. 4, pp. 245–9.

Pynchon, Thomas. *Gravity's Rainbow.* New York: Viking Press, 1973.

Pynchon, Thomas. *V.* New York: Harper & Row, 1990.

Ranke, Winfried. "Magia naturalis, physique amusante und aufgeklärte Wissenschaft." *Lanterna magica: Lichtbilder aus Menschenwelt und Götterwelt.* Ed. Detlev Hoffmann and Almut Junker. Berlin: Verlag Frölich & Kaufmann, 1982, pp. 11–53.

Rings, Werner. *Die 5. Wand: Das Fernsehen.* Vienna/Düsseldorf: Econ-Verlag, 1962.

Ritter, Johann Wilhelm. *Entdeckungen zur Elektrochemie, Bioelektrochemie und Photochemie.* Ed. Hermann Berg and Klaus Richter. Leipzig: Akademische Verlagsgesellschaft Geest & Portig, 1986.

Rotman, Brian. *Signifying Nothing: The Semiotics of Zero.* New York: St. Martin's Press, 1987.

Sartre, Jean-Paul. *Words.* Trans. Irene Clephane. Hammondsworth: Penguin, 1967.

Schiller, Friedrich. *Sämtliche Werke.* Ed. Eduard von der Hellen. 16 vols. Stuttgart and Berlin: J. G. Cotta, 1904–5, vol. 2.

Schivelbusch, Wolfgang. *The Railway Journey: The Industrialization of Time and Space in the 19th Century.* Berkeley: University of California Press, 1986.

Schlüpmann, Heide. *Unheimlichkeit des Blicks: Das Drama des frühen deutschen Kinos.* Basel/Frankfurt am Main: Stroemfeld/Roter Stern, 1990.

Schmiederer, Ernst. "All Power Proceeds from the Picture." *Infowar: Information, Macht, Krieg.* Ed. Gerfried Stocker and Christine Schöpf. Vienna: Springer, 1998, pp. 203–11.

Schmitz, Emil-Heinz. *Handbuch zur Geschichte der Optik.* 15 vols. Bonn: J. P. Wayenborgh, 1981–95. Vol. 1.

Seeber, Guide. *Filmtechnik.* Halle: W. Knapp, 1925.

Shannon, Claude, and Weaver, Warren. *The Mathematical Theory of Communication.* Urbana, IL: University of Illinois Press, 1949.

Simmering, Klaus. *HDTV – High-Definition Television: Technische, ökonomische und programmliche Aspekte einer neuen Fernsehtechnik.* Bochum: N. Brockmeyer, 1989.

Simon, Gérard. *Le regarde, l'être et l'apparence dans l'optique de l'Antiquité.* Paris: Éditions du Seuil, 1988.

Starobinski, Jean. *1789: The Emblems of Reason*. Trans. Barbara Bray. Charlottesville: University Press of Virginia, 1982.

Suetonious. *The Twelve Caesars*. Trans. Robert Graves. London: Penguin, 1979.

Swift, Jonathan. *Gulliver's Travels*. London: Oxford University Press, 1963.

Todorov, Tzvetan. *The Fantastic: A Structural Approach to a Literary Genre*. Trans. Richard Howard. Cleveland: Press of Case Western Reserve University, 1973.

Tsuji, Shigeru. "Brunelleschi and the Art of the Camera Obscura: The Discovery of Pictorial Perspective." *Art History* 13 (1990): 276–92.

Vasari, Giorgio. *Leben der ausgezeichnetsten Maler, Bildhauer und Baumeister von Cimabue bis zum Jahre 1567*. Ed. Julian Kliemann. Trans. Ludwig Schorn and Ernst Förster. 2 vols. Darmstadt: Wernersche Verlagsgesellschaft, 1983. Vol. 2.

Vertov, Dziga. *Kino-Eye: The Writings of Dziga Vertov*. Ed. Annette Michelson. Trans. Kevin O'Brien. Berkeley: University of California Press, 1984.

Virilio, Paul. *War and Cinema: The Logistics of Perception*. Trans. Patrick Camiller. London: Verso, 1989.

Vogt, Hans. *Die Erfindung des Lichttonfilms*. Munich: Deutsches Museum, 1964.

Wagner, Richard. *The Authentic Librettos of the Wagner Operas*. New York: Crown Publishers, 1938.

Watt, Alan. *Fundamentals of Three-Dimensional Computer Graphics*. Wokingham: Addison-Wesley, 1989.

Wedel, Hasso von. *Die Propagandatruppen der deutschen Wehrmacht*. Neckargemünd: Kurt Vowinckel Verlag, 1962.

Wieszner, Georg Gustav. *Richard Wagner, der Theater-Reformer: Vom Werden des deutschen Nationaltheaters im Geiste des Jahres 1848*. Emsdetten: Verlag Lechte, 1951.

Zglinicki, Friedrich von. *Der Weg des Films: Die Geschichte der Kinematographie und ihrer Vorläufer*. Hildesheim: Olms Presse, 1979.

INDEX

242